Eve

wellcome collection

WELLCOME COLLECTION publishes thought-provoking books exploring health and human experience, in partnership with leading independent publisher Profile Books.

WELLCOME COLLECTION is a free museum and library that aims to challenge how we think and feel about health by connecting science, medicine, life and art, through exhibitions, collections, live programming, and more. It is part of Wellcome, a global charitable foundation that supports science to solve urgent health challenges, with a focus on mental health, infectious diseases and climate.

wellcomecollection.org

Eve

The Disobedient Future of Birth

Claire Horn

First published in Great Britain in 2023 by
PROFILE BOOKS LTD
29 Cloth Fair
London
ECIA 7JQ
www.profilebooks.co.uk

Published in association with Wellcome Collection

wellcome collection

183 Euston Road
London NW1 2BE
www.wellcomecollection.org

10 9 8 7 6 5 4 3 2 1

Typeset in Garamond by MacGuru Ltd
Printed and bound in Great Britain by Clays Ltd, Elcograf S.p.A.

A CIP record for this book can be obtained from the British Library

ISBN: 978 1 78816 689 8
eISBN: 978 1 78283 798 5

Contents

A Note on Language

Throughout these chapters I've endeavoured to use gender-inclusive terms when I talk about pregnancy and reproductive health. To say 'pregnant people' and 'birthing people' is to acknowledge that cis women are not the only people who can become pregnant and require antenatal care, which is a fact.

With regard to writing about race, I've aimed to be as specific as I can wherever possible. However, this book engages with medical and legal data, an area in which specificity is woefully lacking. For example, the US Centers for Disease Control and Prevention (CDC) uses 'American Indian' and 'Alaska Native' as identity categories to track outcomes for neonates (newborns) and pregnant people, but with more than 500 federally recognised tribes in the United States, this does not leave us with precise and self-determined language, nor does it support identifying which specific groups are most impacted.

Finally, as will soon become clear, I am not a scientist: I use the terms 'artificial womb' and 'ectogenesis', although these are not always the words that scientists working on external gestation would ascribe to their work.

1

Of Incubators, Orchids and Artificial Wombs

If you are reading this, the one thing that I know to be true about you is that someone, somewhere, carried you in their body before you were a person. Someone gave birth to you.

As I write this sentence I can feel my own baby move in my uterus. I don't know whether the person who gestated you is your mother or not, but I know they too probably felt the indescribable sensation of your limbs shifting under their skin. And I know that before you were big enough to be felt – before you had even taken the shape of a human baby – their body was your home.

I know that, at some point, they probably wondered when, exactly, you would deign to be born. I feel as though I have been pregnant for more than a year, and simultaneously as if this period of my life is passing in the blink of an eye. Pregnancy time, as a friend put it recently, is all lies. Our due dates are guesstimates and, for the most part, we simply do not know when our babies will decide to show up. They could arrive closer to what is really ten months or they could arrive dangerously far in advance.

If you close your eyes, you can probably conjure up an

image of an incubator. But not so very long ago, in nineteenth-century London during the first years of human incubation, it was strange indeed to witness a baby enclosed in a miniature transparent box. It was much stranger still to learn that after a few weeks of being warmed in this way, a previously struggling infant could emerge ready to be reunited with its parents.

Before the introduction of the first modern incubators in Europe in the late 1880s, mothers and midwives practised the common-sense principle that an ailing baby might be helped by keeping it swaddled and warm. However, the incubators that arrived at the turn of the century were thrilling to a public that could not fathom the survival of an infant in this space of air, metal and glass, between its mother's womb and the world.

The apocryphal story holds that the French physician Stéphane Tarnier visited the Paris Zoo with his mind on the high death rates among newborns in the wards of the Paris Maternity Hospital. Watching how chicks thrived in a warming contraption fashioned by the zookeeper Odile Martin, Tarnier asked whether it might be possible to craft a similar kind of structure for infants. Martin obliged and the *couveuse*, or 'brooding-hen' incubator, made its debut. While Tarnier's incubators would begin in a hospital setting, they soon became a fixture of a much more unexpected locale: the fairground. In 1896 the paediatrician Alexandre Lion and the self-proclaimed physician Martin Couney would open the *Kinderbrutenstalt* (child hatchery) to eager droves of spectators at the Great Industrial Exposition in Berlin. Couney would eventually return home to the United States, where he established a permanent 'incubator baby show' at Coney Island's Luna amusement park in 1903. When the first show was launched in Britain at the Victorian Era Exhibition at Earls Court in 1897, the public was so enraptured that a song about breeding a nation 'by means of incubation' became a swift hit that summer.

Contemporary commentators, divided between horror and delight, fed the craze. A rumour swirled that it had become possible to grow babies like orchids in a hothouse: treat them with light, heat and a safe corral and up they sprouted. *The Graphic* magazine published an image of a well-heeled crowd gathered behind a velvet rope, leaning keenly towards the peculiar glass-fronted boxes. Palm trees add a verdant dimension to an otherwise mechanical picture. Nurses stand in neat rows, seemingly poised for action. The small clocks pinned to their white aprons remind the viewer that these are not people at leisure, but professionals at work. Below, the caption reads: 'An Artificial Foster-Mother'. This was the root of what made the technology so enchanting. It seemed possible, both to the day-trippers who marvelled at the babies in their warming boxes, and to the physicians who managed them, that infants might soon be grown outside the human body.

A full-term pregnancy is forty weeks, and Tarnier's incubators would likely have held infants not less than thirty-eight weeks old. Yet the physician boasted that he was on the cusp of enabling almost the entire latter half of gestation to occur through his technology. Despite the caveats of other emerging experts on the incubator, commentators in medical publications like *The Lancet* and the *British Medical Journal* were quick to believe that an artificial womb had arrived. One contributor argued that the only change that might make the technology more precisely like a human uterus would be if infants were suspended in artificial amniotic fluid. But, another author conceded, this could cause the baby to drown, so the innovative use of warm air in glass was the closest possible imitation. With this feat already accomplished, surely it wouldn't be long before the entire process of gestation could be facilitated through these means. The Victorians may have been comically optimistic in their certainty that artificial wombs were just around the corner, but in the twenty-first century we have finally arrived at the brink of this peculiar dream becoming plausible.

In 2017 researchers at the Children's Hospital of Philadelphia (CHOP) released news of successful animal trials of the first partial artificial womb, a platform they dubbed 'the bio-bag'. They had achieved what even the most confident physicians of the 1890s believed to be impossible: they had re-created the liquid environment of the uterus. Any baby born before thirty-seven weeks is considered premature, with any birth before thirty-two weeks considered very preterm. From twenty-eight weeks, in a well-equipped hospital, an infant has a good chance of survival. While contemporary technologies can be used to assist extremely preterm babies born as early as twenty-two weeks, mortality remains high. The best that can currently be done for these infants is the provision of emergency care to treat the complications of being born before their organs have developed sufficiently to function in the outside world. With an estimated survival rate of just 10 per cent for babies born at twenty-two weeks, approximately one-third of those who live suffer significant health issues.

The success of the bio-bag animal trials raised the possibility that these health complications could be prevented, and that a neonate born nearly four months before their due date could recover in good health. An extremely premature lamb fetus was placed inside a translucent polyurethane bag and was buoyed by artificial amniotic fluid. Like the fluids that surround a baby in a pregnant person's body, this synthetic liquid delivered nutrients to the neonate. After much trial and error, researchers were able to use an external pump that fed oxygen into the bio-bag and flushed out toxins to create a workable approximation of the placenta, the extraordinary organ that grows in pregnancy to connect the fetus to the uterus. The technology ultimately allowed scientists to successfully gestate lamb fetuses from the equivalent of approximately twenty-two to twenty-four weeks in a human, until they were fully developed (at the equivalent of around twenty-eight weeks) and could be extracted in good health. In 2019 the group announced a second round of promising animal trials. The process of securing Food and Drug Administration (FDA) approval for trials with human fetuses is now under way, with hopes that this work might begin within the next few years.

Meanwhile, as of 2022, a team working between Japan and Australia has completed two animal trials of a similar platform, which they are calling 'Ex-vivo Uterine Environment Therapy', or EVE. Unlike the boastful claims of Stéphane Tarnier, the group has been careful to caveat that they have no intention of 'replacing' human pregnancy. Of course their emphasis that this project is in no way intended to facilitate gestation outside the body is somewhat undercut by the curious choice to name it after the biblical first woman.

Their research has shown promise with animal fetuses at an even lower gestational threshold and birth weight than those in the bio-bag experiments, with the intention to treat human babies born as early as twenty-one weeks – just shy of halfway through a full-term pregnancy. And in the Netherlands in autumn 2019 a multidisciplinary research team announced plans to create their own partial artificial womb within five years. Using strikingly lifelike 3D-printed model neonates equipped with sensors, and replicating features such as maternal heartbeat sounds, the team plans to create a technology that can not only gestate extremely premature babies, but also track the specific needs of a given infant and readjust the conditions accordingly.

These projects are distinct from one another in experimental design. They share the potential, however, to revolutionise what is possible in the care of extremely preterm babies. Existing forms of neonatal care are emergency interventions. The baby is given treatments to stave off the effects of being born with significantly underdeveloped organs. The artificial womb, in contrast, extends the period of gestation to prevent these complications from arising to begin with. If it works, it will enable the infant to keep growing as though it had not yet been born. And with scientists anticipating human trials within the next few years, artificial-womb technology is no longer purely speculative. Researchers are finally on the cusp of achieving what Stéphane Tarnier could only imagine 140 years ago. Victorians were impressed by the sight of almost full-term babies in glass boxes – imagine how transfixed they would have been by the research now in progress. The 'before' and 'after' images released by the bio-bag team were eerie and briefly ubiquitous. In the first, a floating, pink-skinned, wrinkled lamb fetus sleeps adrift in a transparent bag. In the second, it has grown soft white wool and its body presses against the plastic surface, waiting to be born. These pictures evoke much the same reaction that people once felt when they first encountered incubators: the curious sensation of peering into the future.

It isn't just studies in neonatology that have brought us closer than ever before to achieving artificial gestation. The development of *in vitro* fertilisation in the 1970s led to years of ethical debate, culminating in the adoption of the '14-day rule' for embryo research as a legal regulation in twelve countries including the UK, and as a strict scientific guideline in at least five countries including the US. The rule has meant that scientists could face sanctions for growing human embryos in a laboratory setting beyond two weeks. For approximately forty years there was no cause to question this limit. Despite their best efforts, no scientist had been able to grow embryos beyond nine days. But in 2016 embryologists at Cambridge and Rockefeller universities respectively reached thirteen days, ending their experiments only to avoid violating the 14-day rule. This was a remarkable feat. Before this breakthrough, scientists had assumed that after around seven days – the point when an embryo would normally implant in a pregnant person's uterus – it would need feedback from their tissue to continue growing. Instead, within a nourishing lab-created culture, the cells were able to implant in a dish and continue to organise themselves, suggesting that embryos might be able to grow outside the human body much longer than previously believed.

In May 2021 a researcher at Weizmann Institute of Science in Israel made an even more jaw-dropping announcement. After seven years of study he and his team created an artificial womb in which they successfully grew mice from embryos into fully formed fetuses. Each embryo was placed in a spinning, liquid-filled bottle, carefully modulated to ensure optimal nutrients and temperature. Mouse gestation is substantively shorter than gestation in humans: we take approximately 273 days to form fully, while they take nineteen. The results of this experiment were no less extraordinary for this shortened timeline. For the first time in history, an animal was grown from an embryo to a fetus in a laboratory setting. The mice were gestated from day five to day eleven. Next, scientists plan to take them to the full term of nineteen days. And after that? The group ultimately hopes to run the experiment with human embryos.

When the Cambridge and Rockefeller researchers first hit the 14-day limit, they kicked off a heated debate in the scientific and bioethical communities over whether it was time to revisit the limit. Some scientists argued that now that we had the ability to gestate a human embryo in culture for fourteen days, we should go further. After all, we know more about the bottom of the ocean and what is happening in outer space than we do about the earliest stages of development. Others insisted that there was still plenty to study in the first two weeks of growth, and it was a slippery slope to go beyond this point. And many took the practical view that there was a middle path: that we might be able to continue research without creating new ethical quandaries, but public opinion needed to be part of the conversation.

In May 2021 these years of deliberation culminated in the International Society for Stem Cell Research's release of updated guidelines advising that the 14-day rule be dropped. In effect, this marked the beginning of a sea change. The ISSCR is the largest global membership body for stem-cell researchers. And in countries like the US, where the 14-day rule is a strict scientific framework but is not enforced through law, the ISSCR's guidelines inform research. This does not mean that scientists will suddenly be free to cultivate embryos for as long as they choose, with reckless abandon. It does mean, though, that how much further we might go with embryo growth outside the body is now an open question.

These respective developments in neonatology and embryology have brought us to an unprecedented moment. We are five to ten years from the achievement of a partial artificial womb for humans, based on current estimates. Which means that not so far down the road, we will probably be able to sustain a fetus outside a human body for nearly half of its gestation. If, as seems more and more possible, one day the growth of embryos in a laboratory and the maintenance of infants in a neonatal ward meet in the middle, we will achieve full ectogenesis: external gestation. Babies might be gestated, from conception to birth, without ever being carried in a person's uterus. While growing a human from the embryonic stage through to full term may sound futuristic – impossible even – it is closer than ever before.

After announcing their animal trials in 2017, the research team behind the bio-bag immediately found themselves caught up in a media buzz. You can see the dismay on Emily Partridge and Alan Flake's faces in some of their early interviews, and who can blame them? There they were, ready to explain that their technology, if successful, could drastically improve prospects for extremely premature babies and their parents. Yet interviewers were generally less interested in asking about, for instance, what went into the artificial amniotic fluid they had created, and more interested in whether they intended to grow babies from conception. Was this, eager journalists wanted to know, the beginning of *Brave New World*?

Brave New World was shorthand for the kind of bleak future with ectogenesis that we might all hope to avoid – one where infants are generated in jars, guarded against any kind of loving relationship and destined to become adults who are thoroughly brainwashed subjects of the state. In Huxley's world, artificial wombs are symbolic of the worst aspects of humanity. The CHOP research team has been clear from the beginning that they have no intention of contributing to ectogenesis. 'No one,' Partridge commented back in 2017, 'is trying to do that.' Describing artificial gestation as 'the stuff of science fiction', she was quick to question whether such a feat was even possible. Since releasing their initial study, the group has renamed their project EXTrauterine Environment for Neonatal Development (EXTEND), emphasising the technology's intended purpose as a means of bridging, or extending, the development already facilitated for the extremely preterm baby within the pregnant person's womb.

Magdalena Zernicka-Goetz, the Cambridge scientist who led the first team to report reaching the thirteen-day mark, has also commented that her lab is certainly not trying to pursue ectogenesis, and that to do so would be incredibly scientifically complex. She is, of course, correct. Developmentally, there is a substantive difference between an embryo at thirteen days and a fetus at twenty-three weeks. Just because we have made advancements at either end of gestation does not mean that full ectogenesis is inevitable. Scientists were surprised that the embryos they grew in culture managed to continue developing on their own – for a week longer than expected – but that doesn't necessarily mean there is not some point at which it would become untenable to sustain embryos without implanting them in a person's body. The truth is: we don't know. And while *ex utero* gestation of preterm neonates seems poised to make it possible for infants to be born and survive at earlier and earlier stages, there could well be a limit to how far that threshold can be lowered.

To date, it is considered impossible for a fetus at less than twenty-one weeks' gestation to survive, and each week within the uterus involves new milestones. As researchers working on artificial-womb platforms have been quick to point out, a fetus below approximately twenty-one weeks would likely have veins too small to connect to the lifesaving technology. But even if these scientists do not intend to achieve ectogenesis, others do. The team that announced the successful growth of mice embryos using artificial wombs, though wary of the ethical issues posed by their work, have been explicit that they wish to gestate human embryos into fetuses.

Even aside from whether individual researchers intend to pursue ectogenesis, the last thirty or so years have taught us that scientific innovation can move from futuristic to commonplace with extraordinary speed. Those of us who were kids in the 1990s remember the sudden transition from a handful of classmates having screeching dial-up internet to a smartphone in every hand. Researchers well know that technological breakthroughs often lead to very different ends than they originally intended. Scientific progress frequently outpaces our regulatory systems, and even our imaginations. The truth is that we have been dreaming of artificial-womb technology since rows of warming boxes prompted the rumour that babies could be cultivated like flowers in greenhouses. But now that we are finally reaching the scientific capacity to create an artificial womb, the question is no longer: *Is this innovation possible?* The question is: *Are we ready?*

In the 1970s twenty-five-year-old socialist feminist Shulamith Firestone penned her manifesto. 'Pregnancy,' she wrote, 'is barbaric.'[1] Firestone remarked that as a direct consequence of the dominance of men in scientific research, we could travel to the moon, but we still hadn't found a better way to gestate humans. In 2018, nearly fifty years after Firestone made these claims, I sat in a crowd of geneticists, embryologists and humanities scholars working under the broad umbrella of topics in reproduction. The audience watched in dumbfounded silence as bioethicist Anna Smajdor made a new case for Firestone's forty-year-old argument. The physical consequences of pregnancy and birth ranged from sustained nausea, dizziness and exhaustion, to trauma, permanent injury and death. How was it that we hadn't 'fixed' this yet? Nodding to the research of scientists in attendance, Smajdor made a confident declaration: sexual reproduction was on its way out, and a new era of automated gestation was soon to begin.

As the audience queued for coffee, they debated her proposal. Some women in the group commented that they had enjoyed being pregnant, that carrying a child had been a profoundly challenging but rewarding experience. Others described relentless morning sickness, haemorrhoids, 'feeling like an elephant' and being treated as public property – suddenly every stranger had an opinion on their bodies and behaviour. In weighing their experiences of pregnancy and birth, the perspectives of each of these women had been shaped by the sense that there was simply no other way. What makes Firestone and Smajdor's argument so provocative is their invitation to think beyond this assumption, to ask how our attitudes towards pregnancy might change if, in fact, there *was* another way. Considered in the context of the scientific research of the last several years, Smajdor's prediction about the end of sexual reproduction is not so hard to imagine after all.

Would we end human pregnancy entirely if we could? Could any person of any gender be responsible for gestating a fetus to term? These debates, after Smajdor's speech in 2018, reflect how the very idea of an artificial womb forces us to question our most basic assumptions about human life. This is not just about the possibility of a collective existential crisis. The social and ethical questions raised by both partial artificial wombs and ectogenesis have real-world implications. Developing these technologies will require trials on extremely premature babies. What are the ethics of asking parents to consent to partial artificial-womb treatment? Rates of preterm birth and maternal morbidity and mortality are vastly unequal, with more than 90 per cent of preventable deaths of infants and birthing people occurring in the Global South. The partial artificial wombs that are currently in development are a true game-changer for neonatal care: they could save the lives of countless preterm babies that would otherwise die. But this technology is likely to be extremely costly and to require a substantive infrastructure in order to be used safely.

Whose babies will have access to this treatment? Is there a risk that this technology could increase existing health inequity by improving care for some and not for others? Within wealthy nations like England, which have much lower rates of preterm birth and maternal mortality overall, there is an unconscionable racialised disparity in these health outcomes. As of 2022, Black women in Britain and their babies were four times more likely to die or experience serious medical complications than white women. If artificial-womb technology is available in high-income nations like Britain, would all pregnant people be granted equal access?

And research at the other end of gestation prompts another set of pressing questions. Now that the ISSCR has recommended lifting the 14-day limit, will there be a new global consensus on how long human embryos can be grown in laboratories? Would it be ethically permissible to cultivate a human from the embryonic stage until it became a fetus with fully formed organs, as scientists have already achieved with mice? If a baby was grown through ectogenesis, who would its parents be, and who would be responsible if anything went wrong? Would people be able to choose to use the technology to gestate and, if so, under what circumstances? Would access be limited to people who were unable to carry a pregnancy, or could you simply decide which option you preferred? Even as of 2022, some nations continue to have laws in place that criminalise pregnant people who are believed to be engaging in behaviour that might harm their fetuses. If artificial wombs were widely available, could women perceived to be 'unfit' mothers be coerced into using them? And if a fetus could survive without being dependent on a pregnant person's body, how would that impact upon reproductive rights?

Considered together, the questions that artificial wombs raise for society – for law, medicine and ethics – could have a profound impact on what it means to be human. Today it is a fundamental and uncontested truth that someone gestated you. They navigated the physical, emotional and social ups and downs of being pregnant. They ate and drank and moved, and you reflexively did these things with them. Their heartbeat was your first sound, their uterus was the first place you hiccupped, stretched and spun. This was your first relationship. This was the first person to mediate between you and everything else. What if, alongside all the other things that I cannot possibly know about you, I did not know whether you were gestated by a person or by a technology?

With the partial artificial womb in the immediate future, and progress towards full ectogenesis well under way, we need to begin thinking about the societal impact of these technologies. This is the moment to engage in these difficult conversations. Debates over the uses and dangers of artificial wombs have already begun in academic journals and lecture halls. But conversations that tackle all the complicated questions that ectogenesis raises need to happen in public space. We might reasonably hope that this technology would be welcomed for the way it could benefit the health of neonates and of pregnant people. Many of the loudest voices on the topic of artificial wombs, however, are conservative bioethicists and media commentators who have proposed regressive uses that would undermine the health of pregnant people, not support it.

Some lawyers and legal scholars, for instance, have been arguing for decades that the development of this technology will necessitate the rollback of reproductive rights. One American lawyer pontificated in the late 1970s that after artificial wombs, the law would compel women seeking an abortion to have their fetuses extracted, to continue to grow through ectogenesis instead. The idea that artificial wombs would make it okay for a person seeking abortion to be forced to undergo an operation to extract her fetus and submit to it being brought into the world in a machine is antiquated, cruel and anti-feminist. We could perhaps write off this argument as a relic of a distant past, if not for the fact that since the announcement of the successful animal trials of the first partial artificial wombs in 2017, legal scholars have begun making this very same claim once again. At a conference in 2018 I sat in a stuffy room while a bioethicist explained that these promising innovations meant that, in the not-so-distant future, abortion could be forbidden. This argument was especially unsettling given that, in many countries in the world, people are still struggling for basic access to abortion care. Just months after the final draft of this book was completed in April 2022, the US Supreme Court overturned *Roe v. Wade*, the judgement that had protected a private right to abortion since 1973. Their ruling means that people who find themselves pregnant in anti-abortion states face the threat of forced gestation and birth, or criminalisation for terminating their pregnancies in defiance of unjust laws. Artificial wombs are being researched in a world where people's bare rights to decide whether to continue or end a pregnancy are being stripped away. Those who were not watching the landscape of reproductive rights in the United States were shocked by the fall of *Roe*. But the court's decision followed a decades-long erosion of both access and rights to abortion. The Supreme Court's recent ruling is a stark reminder of what can happen when we look away, or assume that the needle of progress will always move forward. Regressive political actors stand ready to use emerging technologies to undermine our human rights. What kind of grim future would it be if, instead of creating a world where no one is criminalised for trying to control their own reproductive life, we created one where abortion was universally banned and people were forced to have their genetic children gestated against their will?

These are not the kinds of conversations that should be left to conservative bioethicists, legal scholars and researchers alone. When the ISSCR announced in 2021 that they were advising

that the 14-day rule be lifted, they emphasised the importance
of public consultation. And even scientific researchers who are
sceptical of whether full ectogenesis will ever be possible have
acknowledged the importance of discussing the social and
ethical issues presented by artificial wombs. After all, gestation
is one of the few experiences that can be said to impact everyone.
For each of us to exist, someone gave birth to us. If that changes,
so does life as we know it.

Modern though we may be, if the commentaries that have
followed each recent advancement in neonatology and embryology are anything to go by, many of us are just as fascinated
and perplexed by the idea of babies growing in artificial wombs
as those eager viewers of incubator shows at the turn of the
twentieth century. Over the last five years, in the pages of the
Guardian, the BBC, the *Daily Mail*, *The New York Times*, *Discover* and the *New Statesman*, to name but a few, reporters have
speculatively weighed the possibilities of the technology. Alongside images of the bio-bag, headlines have been punctuated by
another clip featuring gynaecologist Guid Oei, lead researcher
on the Dutch *ex utero* gestation project, standing before what
appears to be a giant bouquet of whimsical red balloons inter-
spersed with plastic cords. Despite what is implied in some of
this media coverage, the design is not a working prototype, but
a speculative installation created by Lisa Mandemaker and the
Next Nature Network. These images were eagerly shared by the
press, amid opinion pieces on the future of reproductive rights,
gender equality and human nature. Pictures of the doctor posed
before the floating artificial wombs – like those of the lamb in its
liquid environment – are in notable parallel with *The Graphic*'s
nineteenth-century illustration of the 'Artificial Foster-Mother'.

As the questions put to the bio-bag researchers tell us all too
well, when most people today think of artificial wombs, they

think of *Brave New World*. But there are lesser-known visions of ectogenesis that crop up in sci-fi and fantasy. In Marge Piercy's 1976 novel *Woman on the Edge of Time*, artificial wombs are a tool of empowerment. The technology is a way out of a world where mothers are left to shoulder the trials of pregnancy and the pain of childbirth alone and are subsequently held responsible for anything that happens to their children for the rest of their lives. In the classless, genderless society that Piercy imagines, babies are gestated through ectogenesis and are assigned three parents of any gender who are responsible for 'mothering' them, with the help of the entire community. Since no one is solely responsible for carrying a pregnancy, everyone is responsible for caring for infants once they are born.

No one asks contemporary researchers if their technology might open the door to *Woman on the Edge of Time*. Why is it so much easier for people to imagine a world where artificial wombs lead to dystopian authoritarianism than a feminist utopia of communal child-rearing? Both Huxley and Piercy's visions of a future with artificial wombs are especially interesting in that, while contradictory, their respective speculations are based on the realities of their present day. We expect this of good sci-fi. After all, what really keeps a reader or viewer rapt is a world that bears some resemblance to our own.

Piercy and Huxley began by exploring the question of what might happen if an artificial womb was dropped into their contemporary contexts. Huxley wrote in 1932, in the wake of many years of popular support for eugenics in the UK, and as the Nazis grew in power. From this place he envisioned the worst ways that a totalitarian society might apply an artificial womb to control reproduction and oppress the vulnerable. Piercy, in the 1970s, wrote from an America in the thick of feminist and civil-rights movements. From there, she imagined that in a better kind of society – the society that activists might build – an artificial womb could be used to make the work of care more communal, to undo the association of motherhood with women only. At the precipice of external gestation becoming a reality, we must ask the same question that formed the basis of these two visions of ectogenesis written many decades apart. If we had access to artificial-womb technology, how would we use it?

This book starts from the premise that artificial wombs will only be as innovative as the social context in which they arrive. In an ideal world, partial artificial wombs would be accessible to all pregnant people, to be freely chosen as a means of saving their lives and health and those of their prematurely born babies. And in the further future, ectogenesis would be a tool that people could use to build families of their choosing, regardless of their gender. It would be a means of realising kinship through love and intent, above genetics and designated sex. But we do not live in an ideal world.

Each chapter of this book explores a different facet of how our society needs to change before the introduction of artificial-womb technology. It spans backwards in time to the incubator baby-show craze of the 1890s, and to the origin of the word 'ectogenesis' in a lecture theatre in 1923. And it reaches forward to the scientific future that might be, with babies grown from conception to term through external gestation. Along the way, it dives into the ways in which the laws, policies and institutions that shape our world currently limit the possibilities for ectogenesis and put us at risk of the technology worsening existing inequities and undermining progressive human rights. Ultimately, it traces a path towards the kind of future – beyond the inadequate reality we have created, and beyond the limits of our own imaginations – in which artificial wombs might change humanity for the better, after all.

2

An Artificial Foster-Mother

If you were the parent of an extremely prematurely born baby and doctors informed you that you could choose between a combination of incubation and ventilator treatment or a highly experimental artificial womb, how would you decide what was best for your child? You would be advised about the prognosis and risks associated with each process. Depending on where you lived, you might be told how much each option would cost. You would weigh up all this information and, ultimately, you would take a chance either way.

While artificial wombs are yet to come, the parents of preterm neonates already face the challenge of choosing between different forms of care for their babies. As I write this, I am twenty-six weeks pregnant. Given the nature of my research, I'm acutely aware that every week that my baby remains inside my uterus brings them a little closer to the possibility of being born in good health. I don't believe anyone but a person who has been directly faced with having a baby who requires neonatal intensive care could truly know what this experience is like. But as someone who is pregnant for the first time and anxiously hoping that nothing will go wrong, it doesn't require much imagination to understand how overwhelming that situation would be.

Before a new neonatal technology ever becomes available as an experimental therapy, it must pass through substantial scientific testing and ethical assessment. If my baby came extremely early and I was offered experimental artificial-womb care, I would have questions about previous clinical trials of this technology. I would want to know how many past research subjects had survived, what kinds of complications they faced and whether they had been placed into an artificial womb at a similar gestational stage to my child. In the end, though, my decision would not be based solely on medical evidence and statistics. Faced with a challenging decision, few of us choose by calculating probabilities. My decision would ultimately be instinctive: given the information available, what feels right?

While scientific reasoning is data-driven, both research itself and the ethical and legal guidelines that we set to oversee it are permeated by often-complicated human emotions. Establishing boundaries for how partial artificial wombs are trialled, and clear criteria for their use to treat preterm infants, is necessary for safety and rigour. But it is also necessary because in the situation where this technology is likely to be used first – a situation where someone has just given birth to a neonate whose life is at stake – consent to treatment is anything but straightforward.

Medical consent requires that a person be fully informed of their options, and physically and mentally able to understand those options. It also requires that they agree to a treatment freely, without coercion. Given these conditions, it is unlikely that ethical approval would be granted for an initial artificial-womb trial that involved its use in the immediate aftermath of an extremely preterm birth. Assuming, for instance, that clinical trials were intended to begin with neonates at twenty-two to twenty-four weeks, labour that occurs this prematurely is a traumatic situation for both the birthing person and their baby. Given the enormity of labour's physical and mental toll, someone who has just been through it is not well positioned to consent freely to participating in a clinical study.

Both the EXTEND and EVE research teams have indicated that it is much more likely that initially the technology would be trialled under more controlled circumstances. Emily Partridge, co-lead on the EXTEND project, anticipates that since her team's work so far has involved extracting lamb fetuses from pregnant ewes to continue to gestate in the EXTEND platform, the first stages of the technology's use with humans would involve 'the 50–60% of extreme preterm deliveries that can be anticipated and delivered by caesarean section'.[1] In other words, pregnant people who were flagged as likely to be at risk of extremely preterm birth might be consulted as early as possible to assess whether they were interested in participating in the trials. Inevitably this would remain a challenging decision for any person to make, but it would alleviate the ethical concerns posed by requiring someone to give consent during or after preterm labour.

As of 2022, the EXTEND team is still waiting for approval from the FDA, which regulates clinical trials in the United States, despite beginning the process in 2017. This speaks to the difficulty of determining the terms under which this research might be allowed. Neonatal research in most jurisdictions – including the US, Australia, Japan and the Netherlands, where partial artificial-womb projects are ongoing – requires scientists to follow regulatory guidelines and meet strict conditions for approval of human-subject trials. At a minimum, researchers will need to make a convincing case that their external gestation platforms are likely to offer an improvement over the available alternatives. As Matthew Kemp and Haruo Usuda of the EVE research group have noted, the major challenge is finding a 'first artificial placenta patient' – that is, the first neonatal patient for whom these therapies will be justified because they offer a significantly better chance of survival and a better health prognosis than existing neonatal treatments.[2] Additional research conducted by the EXTEND and EVE teams certainly appears to be oriented towards establishing circumstances under which human testing might be considered permissible and beneficial.

In 2019 the EVE group ran animal trials to explore the efficacy of their technology in a real-world situation. While their initial experiments in 2017 had involved extracting fetuses from healthy sheep, this more recent project examined the technology's efficacy in cases of intrauterine inflammation. Inflammation of the uterus can occur in both human and animal pregnancies, and in humans it often contributes to preterm birth and is linked to adverse outcomes for the neonate. In studies with lambs, the group demonstrated that EVE could help prevent these outcomes and support care for fetuses at an extremely low gestational age and birth weight.

It might be hard to justify the use of artificial wombs for neonates at twenty-six to twenty-seven weeks' gestation, because they would not necessarily serve these babies better than existing treatments like ventilation and incubation. Offered the choice between a tried-and-tested intervention and an experimental one, if my baby came at this stage of my pregnancy and the predicted outcomes were the same, I'd be likely to go with the established technology. But if scientists could demonstrate that, say, a twenty-two-week preterm has a 10 per cent chance of survival without long-term health problems using existing interventions but a 50 per cent chance with an artificial womb, then a much stronger argument might be made. Had I developed intrauterine inflammation at twenty-one weeks and been informed that a clinical trial involving a scheduled Caesarean and transfer to an artificial womb might give my baby a chance, I would have seriously considered participating. It is possible that this is how the first clinical trials of this technology could begin in the future.

The EXTEND research team has similarly begun investigating situations in animal trials that might offer a justification for clinical human experiments. In their most recent study, the group explored the impact on the neurological development of lamb fetuses of gestating in the artificial-womb technology. Because a common side effect of prematurity is damage to brain development, demonstrating that the artificial womb might prevent these complications could help support a case for trials with neonates. If data from animal studies suggested that artificial wombs significantly reduce the rates of neurological problems among extremely preterm babies, then the benefit of the technology could be argued to outweigh the costs of experimental intervention. And as the researchers have reiterated on several occasions, if neonates were struggling after being placed in the artificial womb during these clinical experiments, they could be removed and treated in a more traditional way. Contemporary guidelines stipulate that either the parents or the researchers could elect to switch to standard treatments at any time. No matter how the initial justification for introducing artificial wombs into a clinical setting is founded, though, some risk will inevitably be involved.

Over the last few years the pandemic has made conversations about scientific data, studies and ethics ubiquitous. Suddenly everyone has something to say about whether the regulation of vaccines is fit for purpose. But as both scientists and healthcare practitioners have emphasised again and again, rigorous processes have been developed for approving new medicines, drugs and technologies. And the reason we have come to have strict regulations for human-subject research is that history has shown that individual scientists and physicians should not be permitted to make unilateral decisions about what does, and does not, constitute an acceptable experimental medical practice. In neonatal research, no less than in research on vaccines, the past can tell us precisely why oversight and accountability processes came to be such dominant factors in scientific studies.

Some of the earliest biomedical investigations with preterm babies occurred in environments where ethical guidelines could not have been more absent: at exhibitions, under the gaze of a fair-going public. The image of the incubator as 'Artificial Foster-Mother' in *The Graphic* holds a certain whimsy, but it is also a relic of a time when babies born too soon could effectively either sicken at home or be handed to an over-confident physician for experimental treatment and public display. The early years of the incubator were pure, unregulated trial and error.

In 1837 the German obstetrician Carl Credé created the *Warmwarren*. The technique involved placing the infant inside a central tub that was heated by warm water circulated in an external wall. The *Warmwarren* proved to be both difficult to maintain and riddled with potential safety risks. From a distance of almost 200 years, it is sometimes easy to forget that each experimental subject was somebody's baby. Today if my baby was delivered early I could ask detailed questions about safety and likely outcomes, to determine what kinds of treatments I did or did not want for my child. However, any mother of an infant in those initial years of incubator research would be entirely at the mercy of a probably highly patronising doctor. Do you want your child to live? Well then, let me place him in a heated metal basin and see what happens. When Tarnier's incubator was introduced to the Paris Maternity Hospital in the 1880s, its initial design had several infants suspended together above a hot-water reservoir. Whose children were they? Despite detailed documentation on how these technologies functioned, we don't know what became of the first patients.

Over time, the system was altered so that the incubator held just one infant, surrounded by hot-water bottles that were frequently replaced by a nurse. While these changes didn't necessarily alter mortality rates, they were a step closer to a more modern approach to incubation. Alexandre Lion's patent in the 1890s took Tarnier's model one step further, functioning with as little intervention as possible. Using a pipe flowing with hot water and an automated warming fan, Lion heated the air in his incubator and added a glass top, so that parents and medical staff could watch the infant within.

As the paediatrician and medical historian Jeffrey Baker argues in his book *The Machine in the Nursery*, the incubator marked the beginning of the medical profession understanding premature babies as patients distinct from those born at term. Rather than simply being considered 'weaklings' and included among the high rates of infant mortality overall, this was the first inkling that infants born early might require unique care.[3] This was the origin moment for neonatology as a discipline. Today, at a wealthy, well-resourced hospital, neonatologists would not hesitate to tend to a preterm baby born alive at twenty-six weeks. But in the 1880s the novelty of interventions to save the smallest of newborns left the physicians who championed incubators struggling to elicit support. It was precisely this challenge that led them to take advantage of the era of World's Fairs to promote their new technology.

How might it have felt to be at the *Kinderbrutenstalt*, among spectators who had paid to see Lion's incubators, watching your baby nestled amid glass? Would you be hopeful, relieved that they were in the seemingly capable hands of a doctor, or would you be horrified by the spectacle? Lion's collaborator, Martin Couney, claimed to have trained in France with the obstetrician Pierre Budin, and he elevated incubator baby shows to a whole new level of popularity. Couney would tour through Turin, Italy, the 1900 Paris World Exhibition in Paris, the 1901 Pan-American Exposition in Buffalo and an amusement park display that he dubbed 'the Infantorium' in Omaha, before finally establishing his permanent baby show, which would remain a fixture of Coney Island's Luna Park until 1943. Babies arrived at Coney Island in the arms of mothers and fathers who had been unable to pay for hospital stays and were directed to try Couney instead.

By the time the artificial womb enters clinical experiments with human infants, there will be strong evidence from animal trials to support its benefits. In the early years of incubation, Couney and his peers operated with no oversight, and it was anyone's guess whether a given approach to treatment would work. For all intents and purposes, though, his patients received a level of care that would have been unavailable anywhere else at the time. The infants were kept swaddled and warm, fed by wet-nurses every two hours and monitored on a twenty-four-hour cycle by healthcare professionals who practised meticulous hygiene.

Couney and his colleagues were engaged in a kind of progressive project by embracing neonates as lives worth saving. An initial wave of enthusiasm and curiosity quickly gave way to debates in medical journals over whether aiding babies that were not hardy enough to flourish on their own was wise after all. By rescuing 'weak' infants, some physicians mused, society would become overrun by weak adults, with dire consequences. In a particularly stark example, a Chicago-based doctor produced a 1917 film titled *The Black Stork*, in which a couple that enters a supposedly 'unfit' marriage conceives a child, against medical advice that the infant will be 'defective'. After the baby's birth, the mother allows it to die. The silent film's horrifying tagline, 'Kill Defectives, Save the Nation', also spoke to the doctor's real-world practices. At an inquest during which he was questioned for leaving an infant that he deemed a weakling to starve, he answered, 'I should have been guilty of a graver crime if I had saved this child's life. My crime would have been keeping in existence one of nature's cruellest blunders.'

This is an appalling extreme, but others in the medical profession throughout the early 1900s both privately and publicly supported the position that infants who were born early or struggled at birth might be inherently less worthy of life than those who were robust from the beginning. While, as several journalists decried, displaying babies in incubators rendered them something of a sideshow, physicians and nurses who spent their shifts caring for these premature infants were bucking the trend among many of their peers.

Today intervention to save preterm babies has come to be embraced as a fundamentally good thing, such that artificial wombs to support the care of neonates have been widely celebrated. But it took years of progress for this to be so. As Dawn Raffel notes in her book about Couney, many of the fairs where he brought preemies were concurrently running eugenics exhibitions, promoting the idea that there were 'fit' and 'unfit' marriages, and that societal progression required the former.[4] In Chapter 3 we will see that this history of eugenics is inextricably linked to discussions of artificial-womb technology up to the present day. The incubator babies – small, frail and in need of constant care – grew alongside propaganda to the effect that a baby's ill health, petite size or disability was the consequence of poorly matched parents. According to Claire Prentice, author of *Miracle at Coney Island*, Couney and his team are said to have had 8,000 babies in their care over the years and to have saved 6,500.[5]

With responsibility for infants as early as six weeks premature, that survival rate of 81 per cent is remarkable. To put those numbers into perspective, in the present-day UK, babies born at twenty-seven weeks have an 89 per cent survival rate in a quality neonatal intensive-care unit. While you had to be near enough to one of his exhibitions to rush your infant there soon after birth, Couney accepted babies of all races and social classes and never charged parents. Even after establishing his permanent Luna Park show in 1903 he continued to encourage hospitals to acquire incubators, knowing that his project was not sustainable as a fairground staple forever. Well into the 1930s parents of preterm infants in New York City could come to Couney when they had nowhere else to go.

A century on, in the 2020s, we have accountability procedures for managing human experiments, in part because someone who is desperate and has no other option is not able to give informed consent. Assuming animal trials demonstrate that artificial wombs could significantly improve outcomes for fetuses born at twenty-one to twenty-two weeks, it would still be unethical for researchers to offer this treatment to post-partum parents without providing other options and clear information about the risks. Although Couney was dispensing free care, he was also engaging babies in an unregulated experiment. And although there were no laws or ethics boards to inhibit these activities, there were medical and media commentators who argued – even in the baby-show heyday of the 1890s – that there was something uncouth about these exhibitions. In 1911, after a massive fire tore through Couney's display, the babies were only just saved, and the New York Society for the Prevention of Cruelty of Children began to call for Couney's work to be taken over by hospitals.

At many of these fairgrounds, those who were exhibiting incubator babies (Couney among them) were also participating in a larger practice of treating particular groups as subhuman curiosities rather than as people. The World's Fairs were designed as demonstrations of the scientific prowess and sovereignty of the imperial nations in which they were held. They were sites where human subjects were presented as objects of enquiry, exhibited as a means of colonial propaganda. The historical legacy of the World's Fairs stands as a sobering testimony to the use of research on human subjects as a means of exerting and abusing power. As the reporter Katie Thornton notes of the fairs of this era in the United States, it was common to find:

ethnological villages where Native Americans or people from far away nations would live onsite in stereotyped caricatures of their homes. Some people were literally caged and incarcerated on the grounds with no record of payment. On a lot of midways, there was a despicable willingness to exploit human life for the entertainment of the privileged and charging money to see struggling infants was another manifestation of this unethical practice.[6]

At an 1893 fair in Chicago anthropologists exhibited Indigenous people from around the world in their purported 'habitats'. In 1904, at the Louisiana Purchase, Indigenous Filipinos were similarly put on display by white researchers. Making Indigenous people and premature babies into a spectacle was a performance of imperialist dominance – a gesture to the might of a white empire that could study the people it had conquered as commodities and could experiment with infants who, after all, would otherwise die.

From the moment that baby shows were first embraced by the public, grifters began to create fake incubators. At the Louisiana

31

Purchase, a vendor licence was given to Edward Bayliss, an opportunistic citizen who hired a doctor with little experience to assist him. The babies were kept in overheated and unclean conditions, and were fed cow milk, cereal and egg. A gastro-enteritis epidemic ultimately killed an estimated thirty-nine of these forty-five infants. While we do not know what their fates would have been, had they not been incubated, their deaths were a direct consequence of being subjected to ill care in order to sell tickets.

On hearing of the incident, Couney wrote a furious letter to the *New York Evening Journal*, describing it as the crime of the decade. But he had committed indiscretions of his own. Although he claimed to have received the highest standard of training in Europe, as Claire Prentice uncovered, Couney never actually qualified as a doctor. Unlike Bayliss, whose interest in profits took clear precedent over any concern for the well-being of the preemies featured in his display, Couney's priorities rested with his infant patients. Happily, despite his apparent lack of training, Couney was committed to excellent standards of care and had highly skilled nurses in his employ. Yet we need only look at Bayliss to see that things might have gone very differently. One of these two men with no formal medical qualifications saved thousands of preterm infants. The other's negligence led to the deaths of those entrusted to him. Their diverging stories are testimony to why oversight is crucial to ethics in human-subject research. It would be reassuring to imagine that a singular incident like the deaths of thirty-nine babies at the Louisiana Purchase would be enough to prompt the introduction of scrupulous standards concerning who was permitted to experiment with humans, and under what circumstances. Unfortunately, the world would see many more atrocities before this issue would begin to be addressed.

Fears prompted by past and present incidences of cruel experimentation on people may permeate future conversations about how artificial-womb research is to be regulated. As vaccine hesitancy in the pandemic made plain, the trauma of unethical studies on human beings haunts modern science and medicine. This trauma is compounded by the way that throughout history, and across global examples of human-subject research, these practices have been used to target and dehumanise already marginalised communities.

Where do early incubator experiments fit in this legacy? Individual physicians exploring health interventions for preterm babies may have set out with the intention of saving lives. And, in at least some cases, they may have had the support of the parents of their patients. But whether working in hospitals or at the fairground, they were running human experiments without governance, and without foreknowledge of the costs or benefits of this research.

By the end of the 1930s incubators had fallen from grace on all sides. Some hospitals in Britain, France and the United States, such as Joseph DeLee's Chicago Lying-in Hospital, acquired special incubator wards in the early 1900s, which acted as precursors to the modern neonatal intensive-care unit (NICU). The technology, however, was expensive and unwieldy, and without the broad support of doctors, who were often suspicious of their efficacy in the first place, incubators fell out of favour.

The technology remained sidelined in medicine until the early 1940s, when more and more physicians began to seriously consider the treatment of premature babies. Around the time that New York's Cornell Hospital opened its first premature-baby station in 1943 (the same year that Couney closed his incubator baby display for good), incubation began to be reintroduced to provide both heat and oxygen for preemies. While Lion's original technology had its flaws, the basic principle that the baby should be kept in a womb-like state of warmth still held.

Commentators who had fretted over the incubator shows in the 1920s and 1930s would likely have been relieved that the renewed interest in preterms coincided with a major social shift towards establishing standards for human-subject research. After the Second World War, as the world began to process the knowledge that Nazi scientists had used medical experimentation to torture and kill adults and children in concentration camps, research ethics became an area of international concern. In the aftermath of the Holocaust twenty-three Nazi physicians were tried, precipitating the founding of the Nuremberg Code.

The ten ethical guidelines that make up the code were intended to establish the basics of studies involving humans. They included the stipulation that participants should have the information and legal capacity to give fully informed consent that could be withdrawn at any time. They emphasised the importance of animal trials preceding human ones; proper facilities and preparation for experimentation; and the prevention of unnecessary suffering and harm. And had Couney's work still been in operation, the new precept that people who conduct human-subject research must be scientifically qualified might have given him pause.

The Nuremberg Code was focused on the rights of research participants. Changes were also made in the years following the war that were intended to establish guidance for the activities of physicians and medical researchers. In 1964 the World Medical Association developed the Declaration of Helsinki. This was a document of ethical principles that was codified into law or into medical guidelines in many countries around the world, to lay out the obligations of those undertaking studies on people. In 1982, to support countries in applying the Nuremberg Code and the Declaration of Helsinki in practice, the World Health Organization (WHO) and the Council for International Organizations of Medical Sciences created the International Ethical Guidelines for Biomedical Research Involving Human Subjects. Today the idea that the benefits of human-subject research for society should outweigh the costs is a fundamental principle, but it would take years of establishing governance in research institutions, campaigning for legislation within individual nations and efforts towards international covenants to create a modern foundation for accountability.

From the 1950s to the 1970s numerous attempts were made to try to redress the consequences of preterm birth. As more researchers began turning their focus to preemies, further evidence emerged about why exactly these infants were so fragile. Physicians began to understand that preterms suffered from a plethora of complications caused by their underdeveloped organs. In extremely premature babies, often-serious respiratory conditions resulted from not having enough time in the womb to produce surfactant, the slippery surface that coats our lungs and prevents them from collapsing. Without the ability to expand their lungs outside the liquid environment of the womb, it was not only warmth that these babies needed; it was oxygen. The first efforts at creating mechanical ventilators to help neonates breathe began in the 1960s, laying the groundwork for another key intervention in the care for babies born too soon.

While ethical standards had progressed substantially since the first incubator experiments, neonatal research practices still had a long way to go. Although much media coverage of today's partial artificial wombs represents the technology as being entirely novel, scientists and neonatologists first attempted to create an artificial uterus in the 1950s. The designers of the initial incubators had been informed by the natural environment of the womb. Those researching artificial uteri in the 1950s and 1960s aimed to go one step further by creating artificial amniotic fluid in which the infant could float. If the problem was that the baby's lungs could not inflate in air, they hypothesised, then keeping them in fluid until they had more of a chance to grow might be the solution.

In the 1960s Robert Goodlin, a professor of obstetrics at Stanford University, removed fetuses from women's wombs alive during abortion procedures at ten to eighteen weeks' gestation, and attempted to sustain them through his research. He later acknowledged that at least some of the women on whom he had performed terminations were unaware that he had kept their fetuses for use in his studies. Goodlin would find himself in the unwelcome position of having his work protested by a group of anti-abortion students, an incident that would be documented in the 1974 report of a lengthy US Senate subcommittee investigation into fetal research. In fact, although Goodlin and the student protestors fiercely critiqued one another, their disagreement seemed to turn on the fact that Goodlin had not successfully kept the fetuses alive – the students shared his view that an artificial womb that could 'save' premature infants and fetuses was a worthy goal. They were focused on the potential mistreatment of what they considered to be 'unborn children'. The real ethical issue here, however, was that Goodlin misled, presumably by omission, people who wanted to end their pregnancies, by failing to disclose that he would be running experiments intended to preserve the lives of the fetuses that he removed from them. He did not seek consent.

If contemporary artificial-womb trials are approved for clinical studies, the patients being gestated will be the wanted, prematurely born babies of parents who have given their permission. This is a very different kind of investigation than one with the aim of growing a fetus that has been removed surreptitiously from someone who wanted an abortion. Think about it: in the unlikely event that Goodlin's work had been successful, who exactly would have been responsible for the surviving test subjects? Would the scientist himself be adoptive parent to fetuses that he gestated into babies using his experimental platform?

Unsurprisingly, given that there is still not enough evidence to support human trials of partial artificial wombs in the first years of the 2020s, researchers in the 1960s had limited success using external gestation to keep their subjects alive. These studies are a particularly striking example of the continuation of ethically problematic research practices. But enthusiasm for developing new technologies for preemies meant that similar cases occurred throughout this era. Researchers had moved beyond the amusement park. Yet with neonatology still in its early years, the trial and error of novel methods to care for preterm babies sometimes led to harm. Through the 1940s and 1950s, for instance, doctors puzzled over incidences of retrolental fibroplasia (RLF), a condition that causes blindness. They would later realise that this pattern was not a complication of prematurity, but rather a consequence of their overuse of oxygen as a treatment. As Jeffrey Baker carefully catalogues in his work, 'the imperative to save the lives of small premature infants tended to override the question of whether any of the components, of the many they were introducing, were, in fact, helpful or harmful'.[7] In other words, the costs of experimentation, which included deaths and long-term health consequences, were overlooked because the goal of saving infant lives was considered of great benefit to society, to individual families and to the babies themselves.

How each of us feels about an ethical issue on principle, versus how we feel about it when we apply it to our own families is not necessarily a rational equation. It is all very well for me to say, for instance, that I don't believe individual doctors should be permitted to run untested experimental interventions on preterm babies under any circumstances. The historical record – of preemies dying by accident due to treatments gone wrong – would be on my side. But I also know that if the preterm baby in question was my baby, the actual, tangible baby that I have been carrying for months, then I might feel differently. I might say yes to something that had never been tested, or attempted only once on animals, if I was offered the narrow possibility that it could work. We are creatures driven by emotion, and we often act without logic, even as we think we are being led by reason. Guidelines, and regulations for clinical research with humans, are intended to take the emotion out of it: to set ground rules so that if you find yourself in a challenging medical situation, there are already boundaries for what can and cannot be done. Today, in each of the nations where artificial wombs are under development, many such boundaries are in place.

In the Netherlands, for example, clinical trials are governed by the Central Committee on Research Involving Human Subjects and are subjected to the restrictions of the Medical Research Involving Human Subjects Act. This means that artificial-womb research will be reviewed by an independent medical-research ethics committee that will closely assess whether clinical trials on humans meet the conditions of the Act. This assessment will involve asking questions such as whether research on humans is necessary, whether the possible risks are too extreme and whether it has been designed in a way that provides those who would be involved in trials with enough information to consent voluntarily.

Under European legislation, for medical devices to be studied with humans, it must be very clear that their benefits outweigh their risks. Children are considered a vulnerable population, which means that there are more rigorous standards for trials involving neonates. For this work to go ahead, both parents must give their free approval after being fully informed of the possible risks, in the absence of coercion. There needs to be a demonstrable, direct benefit to neonates. Pain and discomfort must be minimised, and the reviewing committee must have paediatric (or in this case, neonatal) expertise. Parents can withdraw consent at any point, without needing to provide researchers with a justification, and if the child (or in this case, infant) displays behaviour suggesting they are resisting the treatment, it must stop. More so than in trials with adults, researchers need to be able to demonstrate clearly that the interests of the patient will prevail over the interests of science and society.

The Netherlands, Japan, Australia and the United States have all adopted the Good Clinical Practice guidelines as either law or as a medical code of conduct. These guidelines were formed after the International Conference for Harmonisation of Technical Requirements, which was held in the 1990s to attempt to set a world standard for clinical trials involving humans. The thirteen points that arose from this conference emphasise protecting the rights, safety and well-being of human subjects, ascertaining that the benefits of any trial justify the risk and that the intent of clinical trials must be clear and sound. Consent must be voluntary and informed; people conducting the trials must be appropriate experts; and institutional review boards or independent ethics committees must approve the work. These guidelines, and regulation in each of the nations where artificial-womb research is being conducted, repeat and reinforce many of the same principles that were established after Nuremberg and the Declaration of Helsinki. These are the layers of protective measures that are intended to guard against abuse in trials with humans. And they are also intended to protect people from finding themselves in situations where emotion overrides ethics.

Commentary on artificial wombs often jumps from recent animal trials to an imagined future where the technology is widely available, without ever considering the research hurdles to overcome before artificial wombs could be practicable as a treatment for extremely preterm babies. In January 2022 a flurry of activity from tech bros, ignited by comments from Elon Musk, exemplified this phenomenon. Wouldn't it make sense to simply invest in 'synthetic wombs' to make pregnancy easier for women? We will explore the complicated implications of that idea as the book goes on, but for now the very thought that it could be 'simple' to create an artificial womb – even with the backing of Musk and company – overlooks some incredible obstacles along the way. It isn't as though this is a research area that has simply stagnated while lying in wait for Tesla money. Research groups have been working towards some form of external gestation in earnest for decades, with many of these studies never moving beyond animal trials, due to challenges that arose during experimentation.

In the 1970s there was something of a hiatus in research towards artificial wombs. Paediatricians shifted focus to other methods of addressing respiratory challenges in premature babies, such as the use of continuous positive airway pressure (CPAP) to open the lungs. But when researchers began to revisit the idea of a liquid environment to continue a neonate's development outside the womb in the 1980s, they encountered difficulty in gathering sufficient evidence to justify human trials. In Japan a team at Tokyo University worked towards creating an artificial placenta from the 1980s through to the early 2000s, led first by Yoshinori Kuwabara and then Nobuya Unno. Using an 'extracorporeal circuit', the group extracted premature goat fetuses and incubated them within containers holding warmed artificial amniotic fluid. In 1997 they reported that they had successfully kept a goat fetus alive for three weeks using the technology. Despite this achievement, the researchers encountered recurring problems with the animals' heart function and circulatory systems. As Unno put it, his team's goal was not to create an alternative to pregnancy, but to produce 'a total life support system for disabled or extremely sick babies'.[8] In the early years of neonatology, that goal in and of itself might have meant research with infants. In the modern era, the group's successes in animal trials have provided important lessons for contemporary researchers, but they themselves never reached the threshold for clinical studies with human infants.

In the United States, Thomas Shaffer, a neonatal physiologist, worked over the same period towards a different means of re-creating the environment of the womb. Shaffer applied a method that he referred to as 'liquid ventilation', injecting liquid that mimicked artificial amniotic fluid into the lungs of preterm lambs. This was intended to re-create the conditions of breathing inside the womb, enabling the lungs to expand and oxygen to circulate normally. After almost twenty years of research and very promising findings with animals, Shaffer and his team were granted approval by their university and a Philadelphia hospital to attempt their liquid ventilation strategy – with the consent of both parents – on an extremely preterm infant for whom every conventional method of treatment had failed. The baby initially showed signs of improvement, but ultimately did not survive.

There are reasonable ethical questions to ask about this proceeding: is it acceptable to trial a technology on a human for the first time under emergency circumstances? On the other hand, the attempt was backed by two decades of work, multiple animal trials and the reality that the intervention was the baby's best chance at survival. This was the same justification that allowed Shaffer's group to move to clinical trials with preterm human babies at twenty-three weeks. Because these infants were severely ill and had an extremely high likelihood of mortality with existing treatment, there was a strong case to be made that liquid ventilation could offer them a much greater possibility of staying alive. With several of the enrolled infants showing signs of lung improvement, and no adverse reactions that were directly linked to the treatment, the group was able to move to a second set of clinical trials.

In 1996 the second study with thirteen preterm babies between twenty-four and thirty-four weeks' gestation who had severe respiratory conditions resulted in the survival of eight seemingly healthy infants. Again the study could be justified on the grounds that liquid ventilation might offer significantly improved chances of survival and lowered morbidity. These were babies who were assessed by neonatologists to be at high risk of death and were not showing improvement with existing treatments. The trajectory of this work up to and including clinical trials is revealing of the challenges of scientific research and studies with neonates. Even after years of data, animal trials and approval for two clinical human studies with modestly promising results, Shaffer was ultimately unable to fund further experiments. To get to the point where an artificial womb is offered as an experimental treatment, let alone as standard care in a NICU, may take decades. And then there is the question of whether it will work.

For much of the incubator's early years, and well into the era of streamlined neonatal units in hospitals, parents were largely viewed as potential sources of infection to their preterm babies. While it was acknowledged by some physicians, even in the baby-show days, that breastfeeding and proximity to the mother might aid an infant's health, it was not standard practice to treat families as participants in their preemie's care. This began to change in the 1970s, after researchers observed that while there were very high death rates among prematurely born babies in Columbia, infants who were held close to their mothers on a regular basis could flourish. Kangaroo care, or 'skin to skin' contact, where a preterm infant is laid mostly naked on their parent's bare chest, is now standard practice for both preemies and babies born at term. Though technological interventions, from the incubator to the artificial womb, tend to generate the most public excitement, it is perhaps more incredible that being held closely against someone else's skin has true physiological benefits.

Researchers working on artificial wombs are aware of the importance of parental involvement. The EXTEND team reflected from their very first trials with animals that 'parental perception' of their baby being gestated in a bag was likely to be a significant limitation to the project. Features to allow the baby to hear parental heartbeats and voices are intended to go some way towards addressing this discomfort, and they may also have a tangible health impact, too. While babies born at increasingly early stages of gestation, from between twenty-three and twenty-six weeks, can suffer from long-term health challenges, it is also true that many of these preterm infants already survive after treatment in the NICU. Artificial wombs may more closely resemble the environment of the womb than existing strategies for preemie care, but we don't really know what the possible consequences of extending gestation outside the body (even for a few weeks) may be, because no one has tried it yet. A reciprocal exchange happens between a pregnant person and a fetus throughout pregnancy. What are the aspects of that exchange that simply cannot be replicated?

Approaching the last trimester of my own pregnancy, I notice something new every week: how the baby kicks when my dog barks in a particularly hysterical way; the baby's hiccups hours after I eat something spicy; a back-stretch I do that makes them wiggle every time. These small things make me wonder how many other minutiae of my body and the baby's are meaningful, beyond the importance that I as a not-yet parent attribute to them.

What are the ethics of trying to extend gestation outside human bodies? When it comes to projects oriented towards neonatal life support, we know that the standards for clinical trials will be high. We also know that many people are likely to support innovations with the goal of saving preterm babies. But what about full ectogenesis? For this to happen – that is, for a baby to be gestated outside the body, from the embryonic stage through to birth – we would need research at either end of human gestation essentially to meet in the middle. Scientists growing embryos in culture would have to extend their experiments to enable the embryo to grow for as long as possible, and researchers working to sustain extremely preterm neonates would have to continue to lower the threshold at which these neonates could survive. Eventually we would either have the technology to enable an embryo to grow in culture until it could be safely transferred to continue to develop in an artificial womb, or we would have one platform that could sustain gestation from implantation through to birth. However, there are quite substantial barriers, both scientific and ethical, that need to be addressed for this to become possible. How long, for instance, would we allow a researcher to grow an embryo in a lab? Would it be acceptable if it continued to grow to the fetal stage? Who would parent it, if it survived to become a baby?

Couney would have been thrilled to learn that advances to save prematurely born infants at earlier and earlier gestational ages are, for the most part, wholeheartedly embraced in the modern era. Progress in embryo research, on the other hand, has met quite a different response. Given that so much of the concern over embryo research is purportedly down to protecting 'potential life', it might seem like a contradiction that studies on neonates would generate less controversy, since they require experimentation with born babies. Yet partial artificial-womb research has the clear intention of preserving life by supporting infants born too soon. In contrast, growing embryos in culture has benefits like increasing our knowledge of infertility, but also involves cultivating embryos that may then be discarded.

If everyone agreed that an embryo had no more value than any other clump of cells, or conversely that an embryo was sacred, setting limits on this kind of research would be straightforward. But the question of when life begins is extremely fraught. The regulation of embryo research is perhaps the best example of how emotions feed into the laws and guidelines that we create. How do you write regulation to govern something about which people hold such polarising views? For some people (myself included), an embryo is no different than any other cluster of cells. In the early weeks of my pregnancy I was nauseated and tired, but I didn't think of the embryo inside me as a baby. I attached that meaning to it over time. If the embryo existed outside my body from the beginning, in a dish, my emotional response to it would have been shaped by its use. If the embryo was intended only for research purposes, it is unlikely that I would have given it a second thought: to me, it would be tissue that was no longer mine. If it was intended to be implanted into me to attempt a pregnancy, however, I might have felt strongly invested in it. Embryos often hold significant importance for people who struggle with their fertility and go through substantial physical and emotional hurdles to try to become parents with technological assistance. Scientists who cultivate embryos in culture in laboratory settings are aware of the importance of studying these cells to learn more about infertility, and to make future contributions to preventing miscarriage and congenital conditions.

In contrast, some very vocal political actors perceive the embryo as an 'unborn child'. People often assume that the Catholic Church has perpetually marked conception as the beginning of human life. For centuries, though, the Church effectively permitted abortion that occurred before 'quickening' – the point at around eighteen to twenty weeks when a person feels a fetus move in their womb. Aside from a brief and unsuccessful attempt to curb abortions in the 1500s, it wasn't until 1869 that the Vatican issued a papal bull establishing that any termination from conception onwards would result in excommunication. While many in the Church might argue that life begins at conception, it is really the 'potential' for life that is at issue in Church doctrine. It is with reference to the human embryo's purported 'potential' for life that the Vatican has consistently issued directives in opposition to embryo research. In some nations, including the United States, an almost fanatical investment in 'potential' life has meant that anti-abortion campaigners have continuously used emotive language to advocate against work with embryos, insinuating that terminating an embryo used for research is akin to murder. Not all religious traditions approach this kind of research in the same way. In both Judaism and Islamic theology, personhood, or ensoulment, is understood to occur later in gestation, not at conception. In some Jewish traditions life does not begin until birth.

What makes the regulation of partial ectogenesis in embryology particularly complex is that even people who do not attribute embryos with life may feel that there should still be limitations on the breadth and nature of this work. For instance, I would feel ethically comfortable donating an embryo to be cultivated in a lab for a limited period to contribute to a study on miscarriage. I would not feel comfortable donating an embryo that researchers intended to grow in a lab for as long as they possibly could, simply to see how far they might go. That distinction is not necessarily reasonable: if I don't think of an embryo as having inherent meaning, why should I care how it is used? Again, we are irrational beings. Science may be driven by facts, but the research and legal restrictions that we establish to govern scientific practices are very much driven by human desires, fears, contrasting understandings of ethics and gut feelings. What is the right path, when some people think no experimentation should be permitted, others would set limits and still others might simply wish to eschew boundaries altogether?

The questions concerning how to account for the varied ideas people have about embryos, and whose ideas and interests really need to be recognised through regulation, have posed substantial challenges for scientists and lawmakers since the 1970s. In 1969 Robert Edwards and Patrick Steptoe garnered frenzied news coverage when their successful attempt to fertilise human oocytes (eggs) *in vitro* was published in *Nature*. Alongside their collaborator Jean Purdy, they had used donated sperm and eggs to create an embryo in a dish. Scientists had been trying to fertilise oocytes in laboratory settings for years as a treatment for infertility. From the late 1960s through to the mid-1970s more people registered to participate in Steptoe and Edwards's experimental *in vitro* fertilisation trials than the scientists could keep up with. For women who had been struggling to conceive, the possibility that these researchers might be able to provide a solution would have seemed miraculous. What meaning – if any – did the volunteer participants attribute to their embryos? Of hundreds of women enrolled in the trials, only five would become pregnant, and only two of those pregnancies would make it to term.

In 1978 baby Louise Brown became the first person to be born through IVF. Excitement and anxiety over 'test-tube babies' soon reached fever pitch. The scientists and Brown's parents alike received swathes of mail, sent by everyone from hopeful would-be patients to religious fanatics who were angry over what they perceived to be an irresponsible and unnatural step towards ectogenesis.[9] The UK's Medical Research Council, concerned over the political consequences of funding Steptoe and Edwards's work, repeatedly refused their applications. In the US scientists who had hoped to work on IVF were met with ethics hearings. A couple sued Columbia University and Presbyterian Hospital in New York State after the physician who had attempted to perform IVF for them (without undergoing a review process) was halted over ethical concerns before implantation could occur. In India, Subhash Mukhopadhyay, S. K. Bhattacharya and Sunit Mukherjee, who had successfully performed IVF concurrently with Edwards and Steptoe, found themselves blocked from publishing and sharing their findings at conferences, due to political pressure.

For scientific researchers in the 1970s studying human embryos held promise for two reasons. First, it could help them understand miscarriage and infertility and might enable them to treat patients who struggled with both. And second, understanding the origins of human life might also help them trace the links between early development and heritable conditions and diseases. It would be another few decades until scientists were able to culture stem cells from embryos, and until they fully understood the potential of these cells for gaining insight into disorders and illnesses, including cancer, Parkinson's, diabetes and heart disease. In the early 2000s this would introduce a new storm of controversy. But back in the 1970s and 1980s all eyes were on IVF.

Louise Brown's birth to Lesley and John Brown, who had been struggling to have a child for nine years, was meaningful both for people who were trying to conceive and for those who had faced infertility and long since put the possibility of a child behind them. Lesley Brown knew that the birth of her daughter was a remarkable event, not only for her family, but for the rest of the world. The archive of letters, newspaper clippings and congratulatory cards that she kept following Louise's birth is testimony to the varied values ascribed to embryo research. Among accusatory screeds from those criticising IVF on religious grounds, and front-page headlines heralding 'test-tube babies', are cards from women across the globe achingly congratulating Lesley and her husband and sharing stories of their own pregnancy losses.[10]

Suffice to say, the scientists responsible for the first major innovations in the field of IVF were thrust into the spotlight, and politicians and lawmakers were becoming increasingly aware that any decision to permit or forbid this work could be met with a backlash. Ectogenesis loomed large as a touchstone for the dangers of growing embryos in labs. The invocation of 'test-tube babies' that recurred in news coverage invited people to imagine that artificial gestation was already happening, even though the embryos fertilised by Edwards and Steptoe spent very little time in a dish before being transferred to the uteri of potential mothers. Edwards and Steptoe were aware of the spectre that artificial wombs cast on their work. In a 1976 speech prior to the birth of Louise Brown, Edwards had emphasised that they were not attempting to grow babies in a lab. But, he mused, while full ectogenesis was unlikely any time soon, it was true that a brief period of human gestation could now occur outside the body. Partial ectogenesis had been achieved. Other scientific commentators, like the doctors Laurence Karp and Roger Donahue, derided the 'hysterical commentary' on IVF.[11] Media representations of the procedure went to one of two extremes, they noted. Either we were on the cusp of 'glass wombs', which would be a boon for saving the lives of potential children, or we were conducting unethical experiments on 'the unborn'. Both views took away from the issue at hand: helping people who were struggling to conceive.

At the end of the 1970s politicians began to investigate the ethics of embryo research in earnest. In the United States the Ethics Advisory Board was assembled by the Department of Health, Education and Welfare to consider the status of the embryo in 1979. In the United Kingdom in 1982 the government chose the philosopher Dame Mary Warnock to chair a committee composed of seven doctors and scientists of varied religious faiths, as well as eight others, including a theologian, two social workers and two solicitors. The Warnock Committee, as it came to be known, was tasked with assessing whether embryo research was permissible and, if so, how it should be regulated. In 2016 Warnock reflected on the exceptional circumstances that she and her colleagues had found themselves in. Scientists wanted to carry on with their work, but the human embryo as something that could be cultivated in a dish was an entirely new object of research.

While both committees were intended to be reflective of potential stakeholders in embryo research, they weren't necessarily representative of society. Of the sixteen members of the Warnock Committee, for instance, fifteen were white. Their role, though, was not to make unilateral or independent decisions. Instead they had to assess the scientific evidence, hear from community leaders and consider the opinion of any citizen who wished to weigh in. They set out to gather perspectives on the feelings of 'the public' at large. While each committee heard many testimonies against allowing any embryo studies at all, both concluded that there were important scientific and social benefits to this work. Infertility is a source of significant suffering to people who wish to have children but cannot. And the process of *in vitro* fertilisation and embryo transfer presented new possibilities for conception. Both committees also agreed that there needed to be a limit on how long this research could go on, given substantive religious and social opposition. Over the course of hearings with the populace, religious representatives, embryologists, geneticists and academics, the committees respectively arrived at allowing IVF and embryo research, but setting a limit of fourteen days for the period that scientists could keep a human embryo alive outside the womb in culture (*in vitro*).

It was after the Warnock Report's conclusion that twelve countries, including the UK, adopted the 14-day rule into legislation, while others, including the US, applied the rule as a rigid scientific guideline. While scientists in these contexts wouldn't be breaking the law by growing embryos past fourteen days, they could face serious sanctions from their research institutions.

The developmental biologist Anne McLaren was the sole research scientist on the Warnock Committee and presented the scientific case for the 14-day limit. Fourteen days held relevance because it was around this time that the 'primitive streak' appeared – this is the point when cells begin to organise into the outline of what will be the structure of the individual fetus, forming a head-to-tail axis. And it was compelling to both ethics committees as the period when the embryo becomes biologically individuated. It can no longer fuse together or split apart, meaning that the blueprint for an individual entity is set. The map is laid for an embryo to eventually become distinctly you, or me.

Both Warnock and her counterpart on the American commission, bioethicist LeRoy Walters, acknowledged that the choice of fourteen days was somewhat arbitrary. The Ethics Advisory Committee in the US recommended that the limit be *either* fourteen days or the arrival of the primitive streak, which can just as easily occur at thirteen or fifteen days. Yet as Warnock would later recall, drawing a time-marked line in the sand set a clear and enforceable boundary. After the Cambridge and Rockefeller research projects reopened debate over how long embryos should be grown in culture in 2016, a ninety-one-year-old Warnock would comment that her committee's choice came from their understanding that 'law requires the possibility of certainty'.[12] In other words, they wanted to set a precise legal rule that would mitigate any concerns over scientists generating babies in a lab.

This, again, speaks to the way regulation in research is not solely guided by science. Instead it is a tool to try to instil a sense of 'certainty', in the face of public conflict and mistrust. People worried, Warnock noted, that embryos would be kept alive outside the uterus until they were nine months old, and that scientists, having completed their experiment, 'would possibly kill a fully formed baby'.[13] Though the committee emphasised that artificial wombs would not be developed until far in the future, if at all, assuaging these anxieties was part of the rationale for recommending legislation to create a criminal prohibition against exceeding the limit. This recommendation became law with the Human Fertilisation and Embryology Act in 1990, which, though updated in 2008, still governs embryo research in the UK today. The 14-day rule would clearly demarcate the boundaries of how long a human embryo could be grown in a lab, but it didn't necessarily prevent researchers from exploring ectogenesis, nor did it prevent recurring social debates in response to this work.

From the 1980s to the early 2000s several researchers experimented with growing embryos on constructed scaffolding intended to replicate the womb. These were the same years during which scientists renewed investigation into a platform to externally gestate preterm babies. However, while Unno and Shaffer both publicly affirmed that they had no intention of creating an artificial womb, some of these embryologists openly expressed interest in developing ectogenesis.

In 1988 Italian doctor Carlo Bulletti published a paper with his research team called 'Early human pregnancy in vitro utilizing an artificially perfused uterus' in the leading medical journal *Fertility and Sterility*. The paper detailed their efforts at injecting embryos left over from willing IVF patients into donated uteri to see whether they might implant and grow. While the experiment could only be sustained for seven days, the group was hopeful about the evidence it provided that human pregnancy might be replicated in a laboratory setting in its very early stages. By exploring the period preceding and following implantation – the point when an embryo implants in the womb and a pregnancy begins – the researchers hoped to understand more about the causes of miscarriage and infertility. They also tapped into precisely the fears that the 14-day rule was intended to mitigate, in writing: 'the present study was undertaken to obtain the first early human pregnancy in vitro because future complete ectogenesis should not be ruled out'.[14] In context, it isn't as though these researchers were announcing that they planned to replicate the entirety of gestation. The success of their study was limited: cultivating an embryo in a donated uterus for seven days is nowhere near growing an embryo into a fetus and subsequently into an infant. But as the outcry over IVF had so clearly demonstrated, research with embryos was already a fraught subject, and any reference to creating an artificial womb – however far in the future – was inevitably going to prompt a reaction.

Fertility and Sterility is an international medical journal, and a note was included with Bulletti et al.'s paper. The editor stated it was 'the first report of an early human pregnancy attained in an in vitro model', affirmed that the group had been granted ethical approval in Italy and warned against 'serious ethical and legal concerns' that would prevent the experiment from being replicated in the United States. One such concern, of course, was the status of the 14-day limit as a guideline for scientific research in the US. While Bulletti and his colleagues had successfully conducted their experiment and published their findings, they were soon to find that the ethical concerns that the editor suggested would limit such research in the US applied to them, too. In a 2011 retrospective, Bulletti recalled that their work in Italy ceased after the 'political community' mobilised vehement opposition, and questions about ethics were raised. Despite his continued belief that ectogenesis could be achieved and that research on artificial uteri might be beneficial, Bulletti would note that the response to this research demonstrated how difficult it would be to move forward.

In the early 2000s Hung-Ching Liu, an embryologist at Cornell University, would find precisely these barriers blocking her own work on ectogenesis. Liu harvested endometrial cells from mice and grew them into a scaffolding in the form of a mouse uterus. Endometrial cells make up the lining of the uterus, so Liu was essentially working with the body's own mechanisms to create an artificial replication. Once the structure had grown, she cultivated mouse embryos and placed them inside. Liu reported that they were able to grow for a period of seventeen days. In 2003 she grew a mouse fetus from the embryonic stage almost to full term before its death. Having had success with mice, she turned to something bigger. Using human endometrial tissues, she applied her animal-tested method to create scaffolding resembling a human uterus, in which she grew human embryos for ten days.

Interviewed at the Congress of the American Society of Reproductive Medicine, Liu was asked a question heavy with anxiety over ectogenesis: 'Is it [...] science fiction to say maybe in the far future you could have a real breathing embryo and have a child in the laboratory?' Putting aside the fact that embryos cannot breathe and are not children, Liu responded, 'that's my final goal [...] I call it an artificial uterus'. If there was any doubt as to what Liu meant, that was washed away when she reiterated, 'we could possibly have an artificial uterus so then you could grow a baby to term'.[15]

Liu set out to 'help women who had implantation problems'. Like Bulletti, she wanted to discover what went wrong in early pregnancy to lead to losses. For her, growing embryos past the normal point of implantation was research that could help people who struggled to have children. These were not the associations the public had with a scientist declaring that babies might be grown outside the body, though. Dubbed by one reporter 'the nation's premier womb-maker' and hounded with accusatory phone calls and letters, Liu ultimately abandoned her research. Incidentally, in all their speculative coverage of Liu's work, media commentators somewhat missed the mark on the most pressing issue – namely, that Liu's research was not published and thus there were outstanding questions to be asked about its efficacy. In the aftermath, the study never did appear in print.

Given the very different emotive responses generated by studies in embryology versus those in neonatology, it is no surprise that this work prompted substantively more scrutiny than investigations into life-support systems for preemies during the same period. Both sets of scientists were conducting experiments that could feasibly have contributed to ectogenesis: attempting to gestate extremely preterm babies outside the body and attempting to grow embryos externally. But a contributing factor to how the respective work was received was probably also the very different ways in which the researchers talked about what they were doing. Unno and Shaffer were both extremely careful to affirm that they had no intentions of ectogenesis and were sceptical of its plausibility. In contrast, by speaking enthusiastically about the pursuit of artificial wombs, Bulletti and Liu, however unintentionally, triggered the fears that had played such a significant role in generating the 14-day limit.

These studies were bounded not only by regulation, but by what was scientifically possible at the time. Neither Bulletti nor Liu came close to reaching fourteen days. And in the years that followed, there was little debate about ectogenesis, primarily because there was little research to suggest such a thing was plausible. Before 2016, Bulletti was among the few researchers who had published peer-reviewed evidence of growing embryos up to nine days. It wasn't until the remarkable achievement of scientists at Rockefeller and Cambridge universities respectively, cultivating embryos for thirteen days, that the issue was reopened in earnest. These embryos might have continued their process of self-organisation, had the researchers not been required to end the experiment. For the first time, the capacity of researchers to cultivate embryos *in vitro* had outstripped the legal limitations. The question of how long it was permissible to grow embryos outside the body opened anew.

When scientists at the Weizmann Institute announced their hopes of human embryo trials using the mechanical placenta in which they grew mice from embryos into fetuses in 2021, they further stimulated this discussion. In Israel there is currently no limit on how long human embryos may be grown in a laboratory, but the work will still require ethical approval. How, exactly, might these scientists provide sufficient rationale for attempting this experiment? And how, exactly, will the public respond? When the Warnock Committee arrived at their recommendations in the 1980s, the possibility that a scientist could cultivate a human embryo into a fetus had seemed distant indeed. Instead we have reached a new crossroads in determining the limits of such research much sooner than expected.

As discussed in Chapter 1, the ISSCR revised their guidelines in 2021 to lift the 14-day limit. They now recommend that approval be determined on a case-by-case basis through rigorous review. Taking this approach, rather than suggesting a new cut-off, was recommended by the ISSCR on the grounds that research could ultimately outstrip a new limitation once again. This change effectively means that in nations where the limit is implemented through scientific guidelines, as opposed to through legislation, research groups may apply for approval from their institutions to keep cultivating embryos beyond the two-week mark. Universities and research institutions will establish their own standards for what constitutes a sound rationale. The committee stressed that with each proposed piece of research, the costs and benefits of extending the limit should be carefully assessed.

In the US the Rockefeller University group that grew embryos for thirteen days ceased their experiment due to the ISSCR guidance, but could now ask for approval to continue. The ISSCR also advised that countries that currently enforce the limit in law revisit their regulations with public consultation. In these countries it is now politically expedient to do so. After all, a scientist whose work is bounded by regulation in the UK might prefer to bring their study to another nation with more flexible rules.

The question then becomes: what constitutes a strong

justification for growing embryos *in vitro*? That 14-day limit, set in part to assuage anxiety about babies being grown in labs, was intended to communicate a balance between recognising the uses of embryo research and acknowledging the emotional meaning that people attach to embryos. How far could we go without tipping this balance? Scientists who have successfully cultivated embryos up to the limit – including Cambridge-based Magdalena Zernicka-Goetz, and Ali Brivanlou at Rockefeller University – have been clear that while their work could make scientific and social contributions, the question of whether to extend the timeline is one that society, not science, needs to answer.

Among other arguments, those in favour of research beyond fourteen days note the possibility of major breakthroughs in understanding, and perhaps preventing, miscarriage. With an estimated 10–25 per cent of wanted pregnancies ending this way, this research is arguably pressing. Pregnancy loss, like so many issues that primarily impact on women, remains a woefully understudied and poorly understood subject. Learning more about why some people are unable to become, or remain, pregnant could also be incredibly meaningful for those struggling with infertility. Remembering the role that emotions play in instituting regulations and guidelines to govern scientific practice, it is hard to fathom how it would be ethical to weigh objections based on religious ideology equally against research that might allow us to prevent tangible suffering. How many people have had multiple miscarriages with no explanation? How many people have been left feeling alienated by the lack of information about what is happening to their own bodies?

Moving beyond two weeks might also allow for a better understanding of conditions such as spina bifida and diabetes, as well as diseases like cancer and Parkinson's, potentially paving the way for novel treatments in the future. When assessing whether to establish new guidelines, or legislation for how long scientists should be permitted to grow embryos outside the body, we might also consider what is happening around the globe. While many jurisdictions have adopted time limits for embryo research, others, such as Brazil, Israel and France, have not. In other jurisdictions, like China, these limits are in place, but with no clear penalties if they are violated. Do we need a global consensus on research towards ectogenesis? While some might wish to leave the 14-day limit as is where it is already part of existing legislation, not opening this issue for discussion does not necessarily mean that progress towards ectogenesis will not occur. It might mean, however, that this work will only occur in jurisdictions with fewer regulations or accountability measures.

Back in the 1980s the Warnock Committee collaborated to produce their report, but they did not come to a collective consensus. Warnock felt that their disagreements were an important part of the process. If every member of their committee had agreed completely, she commented, they would not have been a very accurate reflection of society. Some of the concerns of religious groups over research on embryos *in vitro* are anti-science and anti-feminist. An agenda that would seek to have all embryos and fetuses protected as persons is one that undermines the rights of pregnant people and refuses to recognise abortion as a medical and social necessity. It is, of course, an agenda that has continually been pushed forward by a conservative, puritanical minority with a staunch political presence in the United States. The Supreme Court's 2022 decision to dismantle the abortion rights established in *Roe v. Wade* sets a new, bleak precedent under which embyros will be treated as legal persons with the consequence of brutal cruelty to the actual people forced to carry unwanted pregnancies or confront possible criminalisation.

The very real potential medical benefits associated with embryo research should not be curbed because some people would have embryos granted greater care than that offered to many living, breathing human children. But it is not solely fringe commentators who have differing views on the ethics and importance of this work.

There are compelling emotional cases to be made for pursuing both embryo growth outside the body and neonatal research on external gestation. One could contribute to studies of miscarriage, and the other could contribute to preventing the trauma of losing a baby born too early. However, we might also reasonably conclude that at some point many people would feel strongly about drawing a line in the sand.

If scientists were able to reach eight weeks (fifty-six days), the embryo would become a fetus. At that point it is still possible that the research might be justified on the grounds of understanding fertility. Would we allow continued experimentation at that stage? If the embryo had been donated by someone struggling to conceive, would it be permissible for researchers to attempt to implant the fetus in their womb? Or in an artificial one?

Given that one neonatal research group is currently anticipating that their technology might be used with fetuses as early as twenty-one weeks, it is possible to imagine that the threshold might eventually shift to twenty and then nineteen weeks, and that each subsequent trial would be justified on the grounds of supporting neonatal survival. But as researchers have noted time and again, there will be very real scientific challenges to going below a certain threshold in neonatal technology. As the scientists who developed the bio-bag have stated, after a certain point, which they predict to be around twenty-one or twenty-two weeks, it might be impossible to connect an artificial-womb platform to the extremely small and delicate veins of the neonate.

Assuming, for a moment, that researchers in embryology could reach that fifty-six-day point when the embryo becomes a fetus, and neonatologists could get to nineteen or twenty weeks (slightly below the current twenty-one to twenty-three-week target), that still leaves months of gestation unaccounted for. We know that technologies can advance quickly. So if it were possible for neonatal researchers to eventually lower the threshold for survival outside the womb even further, and for embryologists to push beyond fifty-six days' growth in a lab, the question becomes whether either of these innovations would be permitted.

Considering the public response to past instances of researchers speaking openly about working towards ectogenesis, and the tensions still surrounding the 14-day rule, it seems likely that at a certain point, people might find experimentation with the external growth of embryos unpalatable. On the other hand, if progress towards ectogenesis were to occur primarily from the direction of neonatal research, then we might not see the same concerns raised. While trials of the technology with neonates will require a very compelling rationale, if partial artificial wombs were to prove safe and effective at increasingly low gestational stages, they might well be welcomed for their potential to support extreme preemies. The research subject would be a wanted, increasingly preterm baby being gestated for a parent.

Inevitably any fetuses and subsequent infants that were involved in trials of ectogenesis, from either direction, would be part of something that had never been done before. What reasoning could there be for the first attempts at growing an infant from conception through to term? Gestating a baby for someone who could not carry a pregnancy?

There are already many ways to make a family that do not involve growing an infant externally. Many people gestate babies without ever being their parent, and many people who never gestate are still parents. Being pregnant does not automatically produce a parental relationship. But every one of us so far has been gestated by a human: and those nine-to-ten months form a relational encounter, a constant feedback-loop, between pregnant person and fetus.

I can't quite fathom what it would be like if the baby that I have carried for almost seven months now had in fact been gestated in an artificial womb. Pregnancy comes with processing vast amounts of information about how the exercise you do, the food and drink you consume, the sleep you get, the anxiety you experience and the medication you take may or may not impact your baby's development. Far from being the most 'natural' thing in the world, you're suddenly encountering an entirely new way of moving through your life – one in which your actions may have physiological consequences not only for you, but for another entity. This is indeed part of what makes the idea of being able to opt in to an artificial womb so compelling to some people. In all the ways, large and small, that we shape babies while they are not yet persons, still entirely dependent upon us, pregnant people are participating in a very real relationship. What happens if that relationship is removed? A life-support system for an infant is decidedly not the same thing as an entirely new form of gestation.

How would the experience of a baby gestated in a technology differ from the experience of human pregnancy? And if we did, ultimately, agree that there were compelling justifications for trialling such a project, what would the consequences be? Rather than speculating about the future, to find some possible answers to these questions, we can travel back to the past, to 1923 and the moment when the word 'ectogenesis' was first coined during a crowded literary gathering in Cambridge.

3

Ectogenesis for a Brave New World

If an artificial womb is ultimately introduced for the purpose of assisting pregnant people and neonates, could we be sure that this is how the technology would be used? We feel differently about how we should limit experimentation with embryos. And we are likely to feel differently about what constitutes a desirable or problematic application of the artificial womb.

Britain's response to the Covid-19 pandemic was an excruciating demonstration of how, at the highest level of government, some lives are valued over others. The repeated claim that no further restrictions were needed, that it was only 'elderly' people and 'people with underlying health conditions' who were becoming seriously ill and dying, exemplified how eugenic ideas permeate modern cultures. Eugenics is any practice by which the state, state actors or others with institutional power seek to kill, limit the reproduction of or allow the deaths of those whose lives they have deemed lesser. It can also include practices by which these same entities encourage or incentivise the reproduction of those who have been systemically categorised as superior. 'Eugenics', which literally means 'good creation', has left a trail of racism, ableism, ageism and annihilation in its wake. It has

63

been mobilised across nations and across time to target particular groups for cruelty. It was the driving principle of the Holocaust, of systematic forced sterilisations of Black and Indigenous people across the US and Canada, of the sterilisation and state-sanctioned murder of disabled people around the world, and of the needless deaths of countless people in nations that elected not to protect those who were vulnerable to the Covid-19 virus.

Why is this relevant to the artificial womb? A technology that enables gestation to occur outside the body could be a means of supporting people who cannot carry a pregnancy. As we will see in Chapter 6, under the right circumstances it could help facilitate a more communal approach to childbearing and rearing. But it could also be a dangerous tool for controlling who is, and who is not, able to reproduce. Looking back in time to the origins of the word 'ectogenesis' offers a window onto the way the legacy of artificial wombs is entangled with eugenics. The idea that this technology would be the ideal means of preventing the reproduction of some members of society, and encouraging that of others, dominated the very first public debates over the ethics of ectogenesis. And this history presses us to think carefully about how far we have come and how much further there is left to go. If this technology was introduced today, could it be used by those in power for eugenic ends? Would contemporary laws and social policies protect against this use?

At Cambridge University in 1923 Charles Kay Ogden's Heretics Society brought together an eclectic set of young writers, philosophers and scientists to discuss the most controversial issues of the day. After living through the First World War and the 1918 flu pandemic – some of them as soldiers, nurses and medics – they returned to university world-weary, anti-authoritarian and with their sights firmly trained on the future. In 1921 Cambridge elected not to incorporate the women's colleges of Newton and Girton, but the Heretics Society was one place where women were welcome to join debates. Charles Ogden himself had invited the socialist feminist Dora Russell to the group before the war and provided a platform for her famously pacifist philosopher husband, Bertrand Russell. Ludwig Wittgenstein, T. E. Hulme, the playwright George Bernard Shaw, the author Virginia Woolf, the nurse and novelist Vera Brittain, and Jane Harrison, the first woman to hold a permanent academic position in the UK, were also among the attendees and speakers. When the society was in its infancy in the 1910s, debates over violence and peace had dominated. In the post-war years, probably as an antidote to the contemporary horrors they had witnessed, the group turned to futurism, keen to speculate on what was to come for science and society. What would their world look like in a hundred years' time? It had to be better than the one they were living in, in which so many had died on battlefields and in hospital wards.

At one meeting in 1923 the idiosyncratic thirty-one-year-old J. B. S. Haldane (geneticist, biologist and writer) gave a speech titled 'Daedalus; Or, Science and the Future'. Before an eager audience, he coined the word 'ectogenesis' and planted the seed of speculation on artificial wombs not only in the imaginations of his peers, but for generations to come. Haldane brought his listeners into a thought-experiment, presenting his ideas in the form of an essay 'read by a rather stupid undergraduate member of this university to his supervisor during his first term 150 years hence'. From the vantage point of 2074, Haldane, as the Cambridge student, recited the history of key scientific discoveries to date. Among the broad subjects covered in the speech was the proposition that, by the 1970s many children in England would no longer be 'of woman born'.[1] In 1951, Haldane prophesied, the first ectogenetic birth would be achieved, after scientific experiments with embryos obtained from a woman who died in a plane crash. After just a few years scientists could 'take an ovary from a woman, and keep it growing in a suitable fluid for as long as twenty years, producing a fresh ovum each month, of which 90 per cent can be fertilized, and the embryos grown success-fully for nine months, and then brought out into the air'. True to the expectations of his Heretics Society audience, Haldane was inspired by actual contemporary research towards isolating and transferring animal embryos.

In the 1890s the zoologist Walter Heape claimed to have suc-cessfully transferred rabbit embryos from one mother to another. And in 1913 the physician Albert Brachet cultured mamma-lian embryos *in vitro*. Haldane included these developments as historical landmarks on the path to ectogenesis. To him, if an animal embryo could be cultured in a lab in the 1910s, it didn't seem at all far-fetched to assume that by the 1970s we would be able to grow human fetuses this way. Haldane went on to predict that despite an outcry from religious leaders, ectogenesis would be universalised. Men and women would find their bodies freed for sexual pleasure without the worry of unplanned pregnancy. Haldane's speculation on this point was endowed with a playful take on what he saw as the logical extreme of changing attitudes towards sexuality – he made a point of describing ectogenesis as the invention that fulfilled the 'the separation of sexual love and reproduction which was begun in the 19th century'. His predictions about the timeline for scientific progress in embryo research were overly optimistic, to say the least, but he did accu-rately anticipate some of the challenges of growing embryos outside the body, and the ethical concerns that have arisen in relation to this work.

'Daedalus; Or, Science and the Future' proved as provoca-tive as Haldane hoped. Charles Kay Ogden would soon expand its reach by publishing it as the first instalment in the *To-day and To-morrow* series. This series of short books written by members and associates of the Heretics Society followed the form of Haldane's initial work, with each author offering their own treatise on the world to come. Most of the contributors were well informed about contemporary scientific research, and most were politically active and outspoken. While the majority aligned themselves with socialist politics, some veered decidedly to the right.

In *To-day and To-morrow* artificial wombs became a way of exploring questions about the future of sexual relationships, what women's social roles should be and whether human gestation is replicable or unique. But beginning with Haldane's initial speech, one question dominated the authors' discussions: should ectogenesis be used to prevent some people from reproducing? Or to encourage reproduction among others? Could ectogenesis be the ultimate means of creating 'better' humans?

When Haldane kicked off this wave of debates over artificial wombs, he presented the technology as a key site for imagining not only the future of gender and human sexuality, but also the future of eugenics. The idea that some humans were inherently more worthy of life than others had existed in some form for many hundreds of years. But in the 1860s the English scientist Francis Galton had begun to argue that everything from physical appearance, to intelligence, to morality was a heritable trait, linked to a person's race and family lineage. He introduced the word 'eugenics' to refer to the process of controlling reproduction to ensure the 'survival of the fittest'.

Galton's ideas emerged from a co-optation of his cousin Charles Darwin's work on natural selection. Darwin had established that while a given species evolves from a common ancestor there will be variations within that species and, over time, variant

67

groups that are best adapted to their environment will survive while others perish. Galton and his peers attempted to apply this idea to humans, hypothesising that in Britain social interventions to protect 'weaker' people were stalling natural selection and preventing humanity from progressing. Galton argued that society could be improved by encouraging the 'fittest' parents to reproduce, while discouraging those deemed least 'fit'. Which groups of people have been cruelly categorised as 'unfit' in different nations, and at different moments in history, is often a telling indication of the priorities of those in control. In Britain in 2020 and 2021, for example, through coded references to 'underlying health conditions' and 'susceptibility' to the Covid-19 virus, the state engaged in the eugenic project of allowing people with disabilities, older people, those who are immunocompromised and those whose work put them in frontline contact with the public (disproportionately those in socio-economically deprived areas and South Asian and Black British people) to die. By taking the stance that each of these groups was in some way biologically predisposed to illness and death ('unfit'), the government attempted to absolve itself, and the broader public, of responsibility for the measures that would have protected them, allowing it to protect its true priorities: the economy and the very wealthy. This was an extension of the same logic engaged in by nineteenth-century eugenicists. Although colonialism and slavery were driven by the willingness of the state to pursue power at all costs, as Angela Saini writes in her book *Superior*, scientists in the 1880s sought a biological basis to justify these atrocities.[2] By systematically classifying people as fundamentally more or less human based on their skin tone, location in the world and social class, eugenicists sought to explain away the brutality of the British Empire.

One of the primary preoccupations of Galton and his peers was the concern that working-class people and new immigrants would reproduce at higher rates than upper classes. Eugenics was perceived as a means of upholding the alleged purity of those who were white Anglo-Saxon, educated and abled, against perceived infringements on that purity. Galton and company cast a wide net around traits they understood to be markers of inferiority: mental illness, 'feeble-mindedness' (a perceived lack of intelligence), physical differences or disabilities, epilepsy or Huntington's disease, alcoholism and drug use, involvement in sex work, a criminal record. Any of these characteristics was considered indicative of being 'unfit'. To substantiate the idea that something must be done about the lower classes 'taking over', eugenicists understood these purportedly undesirable traits and behaviours as hereditary, as genetically associated with racialised immigrants and the working class. As a system of classifying and categorising humans, eugenics was fuelled by white supremacy. By arguing that a person's race and class could make them biologically predisposed to, for instance, illness or criminality, eugenicists attempted to legitimise the treatment of those who were neither white nor wealthy as subhuman.

In the years before the First World War, in both Britain and the United States, eugenics societies arose in affiliation with research institutions and sought to inform the public about their role in ensuring the 'fitness' of the population. These were precisely the groups that organised the fairground contests and information drives displayed alongside incubator baby shows. Beyond these activities, they also campaigned for state-sanctioned policies to discourage people they viewed as inferior from reproducing, through sterilisation, miscegenation laws (laws forbidding interracial marriage) and immigration controls.

As Saini argues, Galton and his peers used scientific language to popularise their claims. In 1907 Galton founded the Eugenics Education Society with Sybil Gotto. The organisation garnered the support of some of the most influential people in Britain at the time. In the 1910s Galton would open the Galton Laboratory for National Eugenics at University College London, establishing an insidious legacy of eugenic research at UCL. In 2018 an inquiry was launched after it was discovered that invitation-only academic meetings of researchers interested in white supremacist theories of race-based differences in intelligence were ongoing. Prominent advocates of eugenics as social policy in the 1910s and 1920s included the Fabian Society founding members Beatrice and Sidney Webb (who worried that the population of upper-class educated Britons was in decline), Alexander Graham Bell (who advocated a ban on marriages among deaf people) and home secretary Winston Churchill, who voiced outspoken support for marriage bans, social segregation and sterilisation for the 'feeble-minded'.

The year of 1905 brought the passage of the Aliens Act, the first significant piece of legislation designed to regulate immigration into Britain. The Aliens Act used coded language to target marginalised immigrants. Restricting the entry of 'criminals', 'paupers' and those with mental illnesses, the law was formulated on fears of the 'dilution' of Anglo-Saxons. Both in the passage of the Aliens Act and in subsequent immigration restrictions to come, a eugenic programme to limit the entry of Eastern European Jews, many of whom arrived in Britain in poverty, underpinned the law.

In 1908 the Eugenics Education Society president James Crichton-Brown went before the Royal Commission on the Care and Control of the Feeble-Minded to advocate that disabled people and those with mental illnesses should undergo compulsory sterilisation, calling them 'our social rubbish' who should be 'swept up and garnered and utilised as far as possible'. Following the findings of the Royal Commission, eugenics campaigners would welcome the passage of the UK Mental Deficiency Act of 1913. The Act updated 1886 legislation to permit the detention and segregation of those deemed to be 'idiots', 'imbeciles' or 'feeble-minded', categories that would allow for the institutionalisation of some 65,000 people at its peak, and which would not be repealed until 1959. While compulsory sterilisation measures were also proposed, they were not adopted. That does not mean that these kinds of measures didn't retain supporters in Britain. In 1931 a Labour MP would renew the call for a sterilisation bill, calling for an end to allowing people 'who are in every way a burden to their parents, a misery to themselves and in my opinion a menace to the social life of the community' to reproduce.[3]

The early 1900s brought waves of such laws to North America. From 1907 to 1931 more than thirty American states passed statutes that allowed forced sterilisation of the 'feeble-minded', impacting at least 20,000 people by 1935. The Canadian provinces of Alberta and British Columbia both passed similar statutes in the late 1920s and early 1930s, with Alberta later opting to remove the requirement for consent from people deemed 'feeble-minded'. Other provinces, despite rejecting sterilisation legislation, practised eugenics in other ways. In Nova Scotia women who were deemed to be fertile but 'unfit' potential mothers were institutionalised.

Numerous American states passed laws that banned or restricted marriage among disabled people and those with some chronic illnesses. The Eugenics Records Office, and its research branch the Eugenics Research Association, as well as the Galton Society in America, brought together like-minded eugenicists to campaign across the country. Eugenics societies in America promoted social-education programmes intended to educate those who were perceived as fit on the importance of marrying and reproducing, sponsoring Fitter Family contests, where families were assessed according to their health and heredity. In 1924 the Johnson–Reed Immigration Act was passed, reinforcing existing racialised immigration policies, effectively banning all immigrants from Asia and significantly limiting immigration from Southern and Eastern Europe. The Act was quite explicitly an effort to retain the presumptive purity of Americans, with one of its champions declaring that its successful passage ensured that 'the racial composition of America at the present time thus is made permanent'.[4]

In the 1920s Harry Laughlin, one of the founders of the American Eugenics Society, observed that while many states had passed sterilisation laws, few people were being sterilised. He wrote what he felt was a stronger, 'model' law for a sterilisation statute, one that would ensure sterilisations would be considered constitutional, should a case arise in court. Laughlin's proposal would lead to the further passage of sterilisation, detention and marriage-restriction statutes across America. It would also act as a blueprint for the 1933 'Law for the Prevention of Offspring with Hereditary Diseases' passed in Nazi Germany, which would facilitate the forced sterilisation of people with disabilities, those with mental illnesses, Roma people and Black Germans.

One of the most infamous moments in the American eugenics movement came with the Virginia Supreme Court case, *Buck v. Bell*, in 1927, which challenged the Virginia Sterilization Act of 1924. Carrie Buck was a young woman who had been institutionalised for 'moral delinquency' after she became pregnant as a result of rape. Her mother Emma had also been institutionalised. In the proceedings – during which Harry Laughlin testified as an expert witness – the court ruled that the Sterilization Act did not undermine rights to due process under the Fourteenth Amendment (namely, the guarantee of civil liberties), and that sterilisation could not be considered cruel and unusual punishment. Voicing the inhumane view at the heart of eugenic ideology, Justice Holmes wrote in the court's decision:

It is better for all the world, if instead of waiting to execute degenerate offspring for crime, or to let them starve for their imbecility, society can prevent those who are manifestly unfit from continuing their kind … Three generations of imbeciles are enough.[5]

This idea, that it would be 'better for all the world' to simply prevent the reproduction of those deemed undesirable, was precisely what drove the passage of forced sterilisation laws across North America. Teenage children of parents who were dependent on federal support would be subjected to hysterectomies and vasectomies, often with the purported consent of a family member who was at risk of losing the resources their family depended on to survive if they refused. Women and girls who had gone to hospitals for other medical procedures would come to understand that their doctors had severed their Fallopian tubes while they were sedated.

In addition to targeting people with disabilities, these policies have continuously had a disproportionate impact on women of colour, especially Black, Indigenous, Mexican American and Puerto Rican women. As Harriet Washington details in her book *Medical Apartheid*, throughout history African Americans have continually been targeted by sterilisation policies.[6] And in Puerto Rico the passage of a law permitting compulsory sterilisation in 1937 was just one aspect of a population-control programme that led to nearly one-third (approximately 160,363) of Puerto Rican women between twenty and forty-nine being sterilised by the time the law was repealed in the 1960s. Many of these women were led to believe that these procedures were reversible.

While sterilisation unsurprisingly looms large in any discussion of eugenics, policies to control reproduction did not only take this form. Miscegenation laws were also expressly eugenic practices, formulated in the belief that by marrying those of another race, white America would slowly degenerate. In Canada the removal of Indigenous children from their families into residential homes was also borne from eugenics, driven by the idea that Indigenous traditions could, and should, be erased by assimilating Indigenous children into white culture. North America and Britain were not unique in their adoption of eugenic policies. The movement that had started in England took hold in nations around the world, with lasting consequences.

This sweep of laws, policies and educational programmes across the globe speaks to a truth that Haldane well knew when he envisaged ectogenesis as a tool for fulfilling eugenics. Had a fully functional artificial womb been available during the heyday of Francis Galton or Harry Laughlin, it is highly likely they would have attempted to engage it in precisely the way Haldane imagined. In the wrong hands – in the hands of a powerful person or government invested in controlling who is, and is not, allowed to reproduce – ectogenesis is a frightening prospect indeed.

In the 1920s J. B. S. Haldane considered himself a 'reform eugenicist'. Unlike many commentators in the 1910s, he criticised state-enforced eugenic policies, calling them 'crude Americanisms'. He played a leading role among geneticists in Britain and the United States who were critical of biological notions of class and race and recognised that nurture and environment shaped people's characteristics alongside heredity. But he still maintained that intelligence was a heritable trait, propagated the idea that people of lower classes were 'breeding' too much and supported measures to encourage upper-class intellectuals to reproduce more. Haldane and his socialist progressive set were aligned with Galton in the 1880s through their protectionist attitude towards whiteness and their belief in the superiority of an erudite elite. After the Second World War Haldane would continue to argue that heritable traits could be controlled and managed in order to enhance the population.

We might welcome the use of artificial wombs for the purposes of neonatal life support, as a treatment to relieve serious health conditions in pregnancy or as an alternative to gestation. Under the direction of a state or institution led by the belief that some people are more deserving of life than others, however, this technology could be used to perpetrate significant harm. The moment Haldane first uttered the word 'ectogenesis' he invoked a link between artificial wombs and eugenics. He imagined the technology as the ultimate means of enforcing the principles of controlling reproduction that had begun with the passage of eugenic laws and policies around the world. It might be comforting to imagine that in the hundred years since then we have broken this link. But how far have we really come? For Haldane, the key benefit of pursuing artificial wombs was their potential use for genetic selection. His narrator, situated fifty-one years into our future, opined:

The small proportion of men and women who are selected as ancestors for the next generation are so undoubtedly superior to the average that the advance in each generation in any single respect, from the increased output of first-class music to the decreased convictions for theft, is very startling. Had it not been for ectogenesis there can be little doubt that civilisation would have collapsed within a measurable time owing to the greater fertility of the less desirable members of the population in almost all countries.[7]

In other words, artificial wombs would ensure that only the 'superior' survived. Other contributors to the *To-day and To-morrow* series would agree that a better future was one in which eugenics had been fully realised.

Anthony Ludovici, writer and devoted follower of the aristocracy, was the first author to respond to Haldane, with his 1924 book *Lysistrata; Or, Woman's Future and the Future Woman*. While at odds with Haldane's progressive views on race and gender, Ludovici agreed on one point: that eugenics would contribute to social progress if applied in the 'right' ways. They did not exactly see eye-to-eye, however, on what the 'right' ways might entail. Much of Ludovici's contribution is a diatribe against the dangers of feminism. Artificial wombs, he agreed, would be a favourable way to control who was permitted to reproduce. But he feared that the young, single working women of his generation foretold a future in which ectogenesis would not be used to further eugenics at all, but rather to support a communal feminist agenda.

Ludovici goes on to describe his dystopia, which some of us might classify as quite the opposite. The daughters of these working women, down through the generations, would continue to fight against the status quo. Eventually their desire for 'extra-corporeal gestation' would be such that 'science will be allowed no rest until a technique is discovered that will meet public demand'. A 'parliament of women' would eventually make it illegal for men to try to impregnate those who preferred to use ectogenesis. Far from Haldane's future, where artificial wombs ensure that only the 'best stock' are allowed to reproduce, Ludovici feared that the working feminist government would require men and the state to pay for however many ectogenetic children women chose. Anticipating that artificial wombs would be expensive, he supposed that Borough Councils might 'grow children for the destitute and the poor' at special centres – to him, a shocking waste of resources. Ludovici's nightmare vision culminates in a reality where humans are ultimately degendered and children are subject to collective care. Of course, as we will see later in the book, artificial wombs do in fact figure in this very same vision for many feminists as a future to be welcomed.

What is striking about Ludovici's screed, however, is that he rejects ectogenesis not because he doesn't want it to be used for eugenics, but because he fears that it will *not* be applied to this end. He argues that the only way to avoid a world rife with weak men and powerful women is to thoroughly apply eugenic practices in his contemporary context. He suggests that the population must be culled through genetic selection: 'abnormal, crippled, defective, incurable, and undesirable people will no longer be allowed to grow up. Their uselessness and their danger as a burden and an eye-sore will be recognized.'[8]

Everyone from those living with illness to people with false teeth, glasses or walking sticks, Ludovici decries, should be prevented from reproducing. He argues that humans, collectively, must alter their values to cease accepting such traits, by choosing mates without them. In so doing, he writes, they might make it so that 'eugenic legislation will become unnecessary and will be anticipated by the taste of people; whereas, if our body-despising values remain intact, Eugenic legislation will always be fighting an up-hill fight'. As Dora Russell would later charge, Ludovici not only held hateful, abhorrent opinions, but also seemed to believe that any rational, educated person would agree with him: a society that endorsed the violent extermination of difference constituted utopia, and a society that endorsed feminism and childcare for all was to be fought and feared.

There are Ludovicis today, too. A doctor whom I won't name, who lives in the Southern states, has a habit of sending me lengthy emails that have a distinctly 'unhinged fear of a feminist utopia' flare. In particular, he likes to share his thoughts about how ectogenesis would be an excellent tool for ensuring the reproduction of the most 'desirable' people, alongside his concerns that women would want to use artificial wombs for foolish reasons, such as to allow them to avoid strain on their bodies during pregnancy. Tempting though it is to write my own tirade in return, particularly while enduring the backaches, sleepless nights, waves of nausea and anxiety, hot flashes, heartburn, breathlessness and general malaise of my third trimester, I have thus far resisted. Ludovici and his modern counterpart are disquieting because they exemplify the ominous ways in which a certain uber-conservative contingent might like to use artificial-womb technology to cause harm. But this is not the most alarming legacy of how ectogenesis is entwined with eugenic thinking. Dora Russell disdainfully referred to Ludovici as 'the author' only when she neatly dismissed what she deemed his paranoid and narrow worldview, and then she moved on. In the same way, I delete my one-way pen pal's emails unopened, rage about them to my partner and carry on with my day.

Far more dangerous and harder to digest are theories about the beneficial uses of artificial wombs for controlling who is allowed to reproduce that come from social progressives we might otherwise find common ground with. Gynaecologist and sexologist Norman Haire contributed a short book to *To-day and To-morrow* called *Hymen, Or the Future of Marriage* in 1927. Like Haldane, he was socially progressive when it came to his views on gender equality, marriage and divorce reform, homosexuality, sex education and reforming moralistic laws around human sexual relations. In his commentary, he calls for free access to birth control and sex education for all, arguing that medical practitioners should provide their patients with open information about sex, but then Haire chillingly pivots from commentary on sexual pleasure to bleak remarks about eugenics. The future, he argues, would see an increase in the state's involvement in producing children, with ectogenesis eventually being used to create 'better babies'. Just as Haldane proposed, he goes on, both the quantity and the quality of these children would be carefully moderated. Once born from their artificial wombs, the children would be looked after by their parents and the state. And, Haire writes, with all this public support available, 'society must have some say as to what sort of children and how many of them it wants'.[9]

Ludovici and Haire in many ways sat at opposite ends of the political spectrum. However, they shared the belief that aristocratic, white, abled people were inherently superior. Haire, in fact, wrote the introduction to Ludovici's *Lysistrata* and noted that they disagreed on most everything, except for the fact that some form of eugenics was in society's best interests. 'Much of our present-day "humanitarianism",' he wrote, 'only results in wasting on the hopelessly unfit money and care which might be spent very profitably on the fit, and in keeping alive those who should never have been born.' In his own book Haire argues that, as knowledge of genetics improves, the bar for fitness to reproduce will be raised. Ultimately ectogenesis could be used to grow 'young embryos of good heredity on both sides'. He comes to a frightening conclusion that speaks to the culture that Coney and his fairground companions were working against, around the very same time. Haire writes that, after ectogenesis, if 'unfit' parents were to have offspring, or if 'any child were born deficient, it would be destroyed at birth, or as soon after birth as its deficiency became unmistakable'.

With few exceptions, Haldane and his peers agreed that it would be possible to gestate a baby outside the human body by the twenty-first century. Had the scientific capacity for ectogenesis been dropped into Haldane's milieu in 1925, he and many of his colleagues would have supported its use for eugenics. It was precisely into our contemporary society that the young intellectuals who contributed to *To-Day and To-Morrow* were reaching. Looking around from our vantage point at a world where many of the scientific innovations they dreamed of exist – a world at the cusp of artificial wombs – would we use this technology differently if we had access to it?

Ludovici and Haire's perspectives on controlling reproduction are equally abhorrent. Ludovici was a Fascist sympathiser and was open in his beliefs that wealthy, educated, white men were fundamentally superior. In his hand-wringing over the very idea of women holding political office, though, he was laughable, even to some of the more conservative of his peers. If it were only fringe commentators like Ludovici, or my anonymous email-writer, who supported controlling people's reproduction, then we might falsely believe that artificial wombs could never be used to perpetuate these kinds of harms in, say, modern-day Britain. But Haire and Haldane's commitment to ectogenesis as a means of ensuring the reproduction of the 'fittest' is testimony to the pervasiveness of eugenics in progressive, leftist circles in the 1920s. To them, policies to manage who was, and was not, permitted to reproduce were not problematic, so long as the appropriate people made these determinations. Haire and Haldane represent a legacy that in many ways is far more insidious than that of Ludovici.

It is not the political alignment of those in power that is the problem. It is the idea that people can be classified and hierarchised as fundamentally better or worse candidates for reproduction, and that reproduction should be controlled on these grounds. The dangerous belief that such practices are permissible, and even beneficial, if they are performed by individuals, states or institutions that qualify themselves as leftist, forward-thinking and well-intentioned is in part what has allowed eugenics to persist into the present day. These ideas speak to how the legacy that Haldane envisioned for artificial wombs could yet be realised, and to how much further we must go to create a world in which ectogenesis would not be used to facilitate harm.

It was as Fascism crept across Europe that Aldous Huxley penned *Brave New World*, cementing an extreme vision of artificial gestation as a dystopic means of state-enforced reproductive control and the hierarchisation of human life. In Huxley's desolate future, the entire process of gestation, from the fertilisation of the embryo through to birth, or 'decanting', occurs through a complicated set of artificial processes in the Central London Hatchery and Conditioning Centre, which is one of several such centres coordinated to do this work around the world. Women contribute ova in exchange for a significant salary bonus. Babies are cultivated based on predetermined criteria to be Gammas, Deltas or Epsilons (the lower-class labouring members of society) or Alphas and Betas (the upper classes). Using 'Bokanovsky's Process', embryos destined to become members of the lower classes are divided multiple times and are deliberately given or denied treatments to enforce their purported genetic disposition for their social class.

When Huxley considered what a totalitarian, eugenic regime would look like, he did so before the revelation of the true extreme to which the Nazis had put eugenics into action. But he was informed by the policies, laws and practices that were in place across Britain, Europe and North America. In 1935 – three years after the publication of his book – the Nuremberg Race Laws set the stage for the Holocaust. Just as statutes had been passed in Europe and North America to justify the sterilisation and segregation of those deemed inferior, so the Nuremberg Laws were designed to categorise Jews, Roma people, LGBTQ people, Black people, disabled people and mixed-race people as less than human. The laws banned intermarriage or the sexual relations of anyone from one of these groups with 'Aryan' Germans and required that people secure a certificate of health fitness to marry. Each of these measures had in part been influenced by statutes passed in the United States and by the writing and recommendations of British eugenicists. In 1939 Hitler passed Action T-4, which authorised euthanising 'the incurable, physically or mentally disabled, emotionally distraught, and the elderly', a programme that continued for years, even beyond the point when it was officially ended. From the late 1930s till the end of the war the Nazis tortured, murdered and incarcerated millions of Jews, Roma and Sinti people, Polish people, Jehovah's Witnesses, LGBTQ people, Black and mixed-race Germans and people with disabilities.

In 1945 the United Nations was established with the intention of charting a path forward. It was hoped that in the act of co-writing and signing the Universal Declaration of Human Rights, nations could commit to basic freedoms and protections to which we might all be considered entitled. Looking back from the vantage point of 2022, it is hard not to feel a degree of cynicism, given the many atrocities in violation of purportedly inviolable rights that have occurred since then. Canada signed its assent while continuing to take children from First Nation communities. The United States agreed that human beings should have equality under the law, even as Jim Crow segregation was ongoing.

The passage of the Universal Declaration in 1948 was an act of optimism. Including thirty entries on which all signees agreed, it remains the basis of international human-rights law today. The intention was to establish the necessities for survival (such as rights to food, housing and asylum) as well as civil and political liberties that all humans needed to thrive. The ideals that undergirded the declaration (equality for all, rights to life and liberty, freedom from torture, the right to found a family) were anathema to eugenics.

Despite the harms that had been caused by eugenic policies and practices in Britain and North America, progressives in Haldane's set did not begin to reassess their outspoken support for eugenics until news of how the Nazis had used such ideas began to spread. But just because they rethought their language choices and public discussions of reproductive control, does not mean that all of these people disavowed the views they had held before the war. Aldous Huxley was critical of the supremacist propaganda promoting the superiority of the 'Nordic' race that had been expounded by both American eugenicists and Nazis. However, as the disability-studies scholar Joanne Woiak puts it, to Huxley, 'eugenics was not a nightmare prospect but rather the best hope for designing a better world if used in the right ways by the right people'.[10] Huxley saw the clear danger in Fascist rule over reproduction, but he did not entirely dismiss the notion that reproduction might beneficially be managed if the decisions were made by people like his intellectual, scientific friends. The language engaged to discuss exerting control over reproduction by liberal commentators today has changed since the 1920s. Callous arguments in favour of sterilisation and allowing only the wealthy and educated to have children would be met with horror in most mainstream circles. Yet the idea that it might be better for both individuals and society if the state intervened in some people's reproductive decision-making remains.

In 2019 I attended a conference where a legal scholar explored whether there were circumstances under which a case might be made in court that using an artificial womb was in the best interests of a child. Put simply, 'the best interests of the child' is a guiding legal principle that requires courts to prioritise a child's interests. In a divorce case, for instance, a judge is ultimately looking to establish what kind of custody arrangement would be best for the child involved. If it can be established clearly that either allowing or preventing an action from occurring is in the best interests of a child, this principle will generally outweigh any other argument.

At this conference the commentator contended that it might be hard to condone artificial wombs. She contemplated, however, several scenarios where a justification to use an artificial womb could be provided that would stand up to legal scrutiny. If a pregnant person experienced complications that made continuing to gestate dangerous for them or the baby, she suggested, a claim could be made that the unknown risk or benefit of the artificial womb would pose a better alternative. What if a pregnant woman was diagnosed with cancer and needed to receive treatment that might be harmful to the baby's growth? It could be in the best interests of the child to continue gestation in an artificial womb. She then went on to propose that if a pregnant person had used drugs or alcohol, the same kind of logic could apply. A case could be made that it was in the best interests of the child to gestate in the 'safer' environment of an artificial womb instead of being carried to term by its mother. In modern writing on artificial wombs, the concept that under some circumstances ectogenesis could be more 'safe' for a fetus than a pregnant person's body has been proposed both by liberal commentators and by those of Ludovici's ilk.

In 2022 a Chinese research project in which scientists created a robot 'nanny' to oversee the growth of embryos into fetuses within artificial wombs made headlines. The work, which is currently limited to experiments with animal embryos, has been described by the researchers as a better means of creating babies. 'Better' because they will be grown in an environment that, as the researchers put it, is potentially 'safer' than a human womb. The unspoken implication here is that the behaviour, consumption habits or activities of at least some pregnant people pose a risk to their fetuses, such that a robot would be a more suitable candidate to oversee gestation.

In a book published in 2000 Dr Richard Gosden pondered that while the human uterus was a 'leaky sieve' to drugs and toxins, an artificial uterus could keep fetuses 'safe'. In a 2006 collection on ectogenesis, bioethicists Scott Gelfand and Gregory Pence respectively suggested that artificial-womb technologies might be useful to 'protect' fetuses from drug- or alcohol-using pregnant women. Pence proposes that 'by raising the fetus in a uniform, stable, drug-free, and controllable environment, the fetus is spared from risks associated with the mother using drugs'.[11] And Gelfand suggests that ectogenesis could protect fetuses not only from alcohol, but from second-hand smoke and a pregnant person's sub-par dietary choices.[12]

Conservative bioethicist Christopher Kaczor writes, somewhat chillingly, that 'partial ectogenesis may someday become less risky than normal gestation, since an artificial womb would not, presumably, get into car crashes, slip and fall, or be assaulted'.[13] The idea that rather than providing a pregnant person with resources to protect them from assault, it would be better to simply extract fetuses from their bodies to gestate in a 'safer' space is deeply disturbing. These arguments render the fetus a person with rights equivalent to a born child and, in doing so, they undermine the rights of the pregnant person, implying that they are no more than an incubator, an 'environment' to be optimised. And each of these arguments also carries forward the legacy of artificial wombs as a means of realising eugenics. In laying out his vision for the future, Haire wrote openly that in the case of an 'unfit' mother, 'either permanent or temporary avoidance of parenthood may be called for in the mother's own interest'. In contemporary imaginings of artificial wombs as 'safer' than human ones, no one directly uses the words 'unfit mothers'. Yet the implication remains clear: some women simply are not fit to carry a child and should therefore permanently or temporarily avoid gestating.

Think about what remains unspoken here. For a baby to be gestated through ectogenesis, a person would have to donate an embryo or else, if they had already begun their pregnancy, have their fetus transferred through a procedure presumably similar to a Caesarean section. How would this occur? Would women who used drugs or alcohol, partook of 'problematic' diets, smoked or experienced abusive relationships simply agree that it was better for everyone if a synthetic womb carried their babies for them? Would they voluntarily undergo an 'extraction' procedure? Or would they be compelled to do so? Let's assume that any use of artificial wombs as an alternative to gestation, where someone's pregnancy was deemed unsafe, would be carefully regulated to ensure they volunteered to undergo a transfer, rather than being coerced. Even then, this would still be a twisted extension of the same problematic logic practised by *To-day and To-morrow* contributors like Haire. In attributing poverty, lack of education and alcohol dependency to genetics rather than social inequality, Haire vested the solution in the wrong place: preventing people who experienced these issues from reproducing. In the same way, contemporary commentators who suggest that artificial wombs could provide a solution for alcohol and drug use in pregnancy understand the pregnant person as inherently suspect, rather than as a human in difficult circumstances. Consequently they come to a misguided conclusion: that the pregnant person's body is the issue, rather than a society that has failed to provide this person with sufficient support and resources. There is another question that remains unspoken in these ideas that ectogenesis could be 'safer' than a person's womb. Who exactly gets to decide what constitutes 'risky' behaviour during pregnancy?

Many people would agree that the suggestion that artificial wombs should replace human reproduction to ensure that only groups deemed 'superior' could reproduce is unacceptable. They might not see a problem, however, with suggesting that a person who smokes, drinks or does not eat 'well' during pregnancy should be offered an artificial womb in which to gestate their child instead. You would think that the nature of my research might have prepared me for the barrage of public opinions that come with being pregnant. I have still been floored, however, by how many people believe it is appropriate to offer commentary on what I should and should not eat or drink, how much weight I should or should not put on, and whether it is a good idea to run while pregnant. None of these comments are made for my benefit. When someone decides to share these thoughts, they are doing so out of a sense of entitlement to my pregnant body, paired with often deeply inaccurate ideas about what constitutes a risk to a fetus. When I arrived for my Covid-19 booster shot, the pharmacist looked me over and informed me that I would most likely get a fever in the next few days, but I should not take paracetamol because that was 'bad for baby'. After I answered that if I had a fever, I would take medication for it, she sighed and told me, 'Yes, a fever isn't good for baby, either.' In pregnancy, almost everything that you do can be construed as potentially harmful to your baby.

Advice for pregnant people varies widely across national guidelines. In Canada we are told to abstain from drinking entirely throughout pregnancy, to avoid certain cheeses and raw foods and to be cautious of various herbal teas. While an obstetrician told me to have no more than one cup of coffee a day, a midwife told me I could have two. The ob-gyn emphasised maintaining a healthy weight as being crucial to my baby's health, but a midwife informed me that my weight was not significant to my care. This contrasts with the National Health Service's Start4Life pregnancy guidance, where you will find intensely condescending commentary about what constitutes an appropriate diet and the chirpy reminder that 'you're eating for you, not for two!' In India, women are told not to eat microwaved or oven-heated food in early pregnancy. In France, drinking wine in moderation (one glass a day, for instance) is not considered to be dangerous. While in the United States, eating sushi is categorised as unsafe, in Japan sushi is considered a healthy part of a pregnant person's intake. Google virtually any substance or device alongside 'pregnancy' for a cascade of comments on how its use, or absence, could cause harm to a fetus. In other words, the rules as to what is safe in pregnancy are culturally-specific, shaped by whether they come from an overly medicalised or person-centred environment, and quite significantly in flux. So if, according to the logic of some bioëthicists, we should use artificial wombs to protect fetuses from 'unsafe' actions by pregnant people, where would we draw the line? And who would get to set that line?

Alcohol consumption in early pregnancy has an association, but not a clear causal relationship, with miscarriage and higher rates of preterm birth. Significant use of alcohol in pregnancy can result in fetal alcohol spectrum disorder (FASD) or fetal alcohol syndrome (FAS), both of which can be devastating for the children who suffer from these conditions and for those who care for them. However, there is no conclusive evidence suggesting that limited alcohol use in pregnancy will cause lasting harm. Many pregnant people do choose to abstain from alcohol throughout pregnancy. But providing advice to pregnant people on their health and that of their babies during gestation is not the same as introducing medical or legal guidelines that would allow others to infringe their bodily autonomy and control their choices.

The trend among bioethicists to think of the artificial womb as a 'controllable environment' through which fetuses could be protected from the harmful habits of pregnant people is a new iteration of a long-standing pattern. Being able to prevent or end unwanted pregnancy through access to abortion and all forms of birth control (including options like tubal ligation and vasectomy) are vital aspects of reproductive freedom. In an ideal world, we would provide the resources and tools to enable people to decide on their own terms whether and when they wish to begin or end a pregnancy, and what kinds of contraceptives work best for them. Being able to decide that you do not wish to become pregnant or continue a pregnancy is very different from someone else deciding for you that you pose a danger to your own baby. When a state or institution determines on someone else's behalf that they should not be permitted to carry their own child, that is eugenics. If a judge were granted the power to rule that it was in the best interests of a child to be extracted from a pregnant person's body, due to that person's behaviour (whether they used alcohol or drugs, were about to undergo cancer treatment or were a victim of abuse), this would be a eugenic and anti-feminist practice. And under what circumstances, exactly, is it likely that someone who faced this ruling would feel they were being given a choice?

Commentators in the present day, like their 1920s counterparts, have offered up this kind of speculation as a thought-experiment, not necessarily arguing that the technology *should* be used in this way, but merely proposing that it *could be*. These hypothetical scenarios, though, are sinister. The idea that some people are potential 'dangers' to their unborn children does not just exist in an imagined future with artificial wombs. It is not an innocuous, philosophical 'what if' concept. We know that value judgements about which kinds of people should be allowed to be pregnant, and granted the autonomy to make decisions about their bodies, their pregnancies and the future of their children, have been used to justify forced sterilisation and the removal of parents from their children for generations. And these ideas do not get applied equally to all pregnant people.

Into the present day, Black and Indigenous women, trans people and those with disabilities have continued to face forced sterilisation and pressure to use birth control at the hands of states and institutions. One of the most pressing challenges of protecting human rights is transforming them from mere principles on paper to something that nations uphold in practice. Sterilisation statutes in Canada and the United States remained in place well into the 1970s. Alberta's sexual sterilisation act, permitting forced sterilisation with the approval of a board, wasn't repealed until 1972, with British Columbia following in 1973. It wasn't until 1986, in the case of *E (Mrs) v. Eve 2*, in which a mother sought permission to force the sterilisation of her adult disabled daughter, that the Prince Edward Island Supreme Court officially ruled against involuntary sterilisation. In South Carolina a sterilisation statute was in place, if not frequently in use, until 1986. In West Virginia, until 2013. *Buck v. Bell*, the Supreme Court case in which the judge infamously decreed, 'Three generations of imbeciles are enough', has never been overturned.

After several high-profile cases there was a greater push in the 1970s to require measures such as the provision of accurate information as to what a sterilisation procedure entailed and ensuring that people who agreed to sterilisation were not coerced through such means as the threat of losing their children. But instances of coercive sterilisation continued in the hundreds of thousands. Historian Jane Lawrence reports that between 1970 and 1976 an estimated 25–50 per cent of Indigenous women in the United States between the ages of fifteen and forty-five were sterilised. Since the Universal Declaration of Human Rights was signed, the United Nations and World Health Organization have repeatedly reiterated that sterilisation without full, informed, voluntary consent is a human-rights violation. The European Court of Human Rights, too, has ruled decisively against these practices. In 2017 the court affirmed that requiring people to be sterilised prior to receiving gender-confirmation surgery was coercive and a significant violation of privacy. While the case dealt with a statute in France, the ruling was also applicable to the twenty-two member countries with similar laws. In 2019 the Inter-American Commission on Human Rights called on Canada and Peru to 'to put an end to the practice of forced sterilizations by adopting legislative and policy measures to prevent and criminalize the forced sterilization of women', to define consent much more comprehensively, to report and investigate sterilisations of Indigenous peoples with transparency, and to train health practitioners on the history of coercive sterilisation.

Incidences of women in US federal prisons being subjected to tubal ligations, and of both men and women being offered reductions on prison sentences in exchange for sterilisation or the use of long-acting contraceptives, continue. In 2020 a whistle-blower reported that hysterectomies were routinely being performed on women detained in Immigration and Customs

Enforcement detention centres. Continuing well into the 1990s, First Nations children in Canada were removed from their families in substantive numbers and placed in residential homes or put up for adoption. Under the directives of the Indian Act, during the 'sixties scoop', thousands of children were taken by provincial agencies and adopted into white families. These events, executed by racist governments under the guise of acting in the alleged best interests of children, are in living memory.

In Canada, Indigenous women continued to be subjected to coerced sterilisation in hospitals as recently as 2018. As Alisa Lombard, a lawyer representing hundreds of litigants across Saskatchewan, has commented, her clients were made to undergo procedures such as tubal ligations in situations where it was impossible to give informed consent. She cites, for instance, examples of consent being demanded from women right after giving birth, while social workers edged towards the door holding their babies. That these atrocious actions are committed in the first place is a consequence of racist stereotypes that frame Indigenous women as likely to engage in 'risky' behaviour during pregnancy, and therefore as unworthy of motherhood.

What do these ongoing practices tell us about the possible harms of how ectogenesis might be used in our future? Scientifically speaking, we have made great strides towards the artificial womb since Haldane first dreamed up the word. Huxley envisioned infants floating, suspended, in liquid environments long before we had even perfected the air-filled chamber of the incubator. The semi-translucent space of the EXTEND platform, in which fetuses as premature as twenty-three weeks will be submerged in carefully measured artificial amniotic fluid, would have captivated Haldane and Huxley. Perhaps it would even have sent them into rapturous musings about how much further we might go. It would be unjust to the generations that have fought against eugenics to imagine that no change has been made since 1923. But if a functional supply of artificial wombs landed in the hands of, say, lawmakers in the United States or government advisors in Britain, could we be certain that they would not be used to take control away from pregnant people?

Clare Murphy, executive director of the British Pregnancy Advisory Service,[14] was not surprised to hear that contemporary commentators had suggested artificial wombs could be used as an 'alternative' to gestation if pregnant people used drugs or alcohol. She tells me that the organisation has struggled in recent years to help people understand why it is a problem to implement systems that monitor pregnant people's use of substances. If a group announced in today's Britain that it wanted to use ectogenesis to ensure that only the 'best' would reproduce, it would be met with protests. But Murphy notes that once a behaviour in pregnancy becomes characterised as 'bad' in the public eye, people are often all too willing to accept infringements on a person's self-determination to prevent that behaviour.

In 2020 the National Institute for Health and Care Excellence (NICE), which creates guidance and policies for the NHS, advised that medical practitioners should ask pregnant people for information about how many alcoholic drinks they consumed throughout pregnancy (even before realising they were pregnant). Healthcare practitioners, including midwives, were further advised to record this information, with or without consent, to retain on the pregnant person's health record and to be added to the health records of their child. The policy proposed suspends the privacy of all pregnant people, supposedly on the grounds of protecting future children from FASD. But very few pregnant people drink to such a degree in pregnancy that FASD occurs. As Murphy points out, most pregnant people either abstain from alcohol or may consume an occasional glass of wine or beer, for which there is no evidence of damage to their babies. And more importantly, the solution to instances where a person does consume significant amounts of alcohol in pregnancy is not covertly monitoring them. It is providing them with patient-centred care. These practices of monitoring are sinister, Murphy notes, for the way 'they erode the trust between the pregnant woman and care provider' and undermine the care provider's ability to communicate about actual risks and how to mitigate them.

This policy reinforces the idea that there are 'good' and 'bad' pregnant people, and that if you do not act to optimise the environment of your uterus, you should not be permitted to control your own reproductive body. In this climate, Murphy muses, it is easy to see how a technology like the artificial womb, which should be used to support and assist pregnant people, could get co-opted to further undermine their autonomy.

The United States is a stark example of what can happen when surveillance of pregnant people is taken to an extreme. Between 1973 and 2005, 400 pregnant people (most of whom were Black, Indigenous, Latina or white women from lower socio-economic backgrounds) were detained on charges of endangering a fetus. These charges overwhelmingly targeted low-income pregnant people dependent on subsidised healthcare. In many of these cases, healthcare practitioners referred pregnant people to the police. Incidences where this occurred included occasions when pregnant people confided in their physicians about past histories of drug use; and occasions when the pregnant person was given a urine test, which they understood to be routine procedure for all pregnant patients, but which was intended to covertly search for evidence of drug or alcohol use. And while many women have been charged under these statutes for substance use in pregnancy, two-thirds of drug-treatment clinics in the US will not admit pregnant women. Where they do, these centres are often sites where pregnant women are referred to the police. Denying care to the mother, and placing her in an extraordinarily stressful situation, is patently not an effective way to protect a future child – if that were, indeed, the goal.

When laws, policies and medical guidelines are produced to monitor pregnant people's activities, enforcement is consistently racialised and classed. Contemporary statutes in several American states allow pregnant people to be detained if they are suspected of consuming drugs or alcohol. A Wisconsin statute, for instance, allows the state to detain people believed to be pregnant who 'demonstrate "habitual lack of self-control" in the use of alcoholic beverages or controlled substances'. Many of these statutes were initially passed to prevent children from being exposed to dangerous substances, but were quickly co-opted to charge pregnant people with endangering 'unborn children'. In some cases, women have been detained for issues unrelated to drugs or alcohol, such as the charge of acting irresponsibly by failing to attend prenatal appointments.[15] These laws disproportionately criminalise low-income Black, Indigenous and Latina women. And to imagine that an artificial womb could be used to essentially take out of their custody, even before birth, the infants of those pregnant people deemed problematic by the state would be to extend this practice into an imagined future. Who, or what entity, would you trust to make decisions about whether a pregnant person's behaviour was dangerous, such that she should use an artificial womb to continue gestating rather than carry her own pregnancy? These kinds of regulations, like existing laws and policies, would be destined to target already minoritised communities.

Given this precedent in the United States, what would happen if ethicists, lawyers and policymakers successfully lobbied to make artificial wombs available where a pregnant person used drugs or alcohol? Consider the case of an Indigenous woman who was arrested and detained in North Dakota under charges of exposing a fetus to toxins. While awaiting trial she sought an abortion, after which the state dropped her case on the grounds that it was no longer relevant. In a scenario where an artificial womb was available, a person in this situation could be informed that charges would be dropped, should she allow the fetus to be extracted to an artificial womb. In this instance the introduction of artificial wombs could raise a very real probability in the US, not of fetuses being forcibly extracted from women's bodies, but, as in the case described above, of a pregnancy being ended by 'choice' by someone who would face substantive criminal charges if they did not comply. Where a pregnant person would need to consent, given that the choice would be between criminalisation and fetal extraction, the situation would remain coercive. Based on existing precedent, this kind of coercion would not be circumscribed by US law, as it would involve no forced bodily procedure and would uphold state statutes targeting drug use in pregnancy.

In Huxley's World State the use of artificial wombs is mandated by law. Unlike the US, in contemporary Britain the introduction of any kind of regulatory framework that enforced the use of this technology is implausible indeed. In Britain very strong legal precedent protects pregnant people from being criminalised for actions towards their fetuses. In one of the most recent attempts to try to bring a lawsuit against a pregnant woman for alcohol use in pregnancy (in 2014), a court decisively ruled that under English law a fetus is not a person. As such, a pregnant person cannot be charged for actions in pregnancy regardless of whether they might cause the fetus harm. So while laws in some parts of the United States might not prohibit women being coerced into using artificial wombs, in Britain it is likely that this sort of action would be inhibited. But controlling reproduction is not simply limited to whether something is permitted in law.

In Britain a two-child benefit limit on Universal Credit has been in place since 2017, which means that recipients of Universal Credit cannot receive any additional support after a third or additional child. The exceptions to this rule are if the person seeking credit has three or more children due to a multiple birth, or if the additional child is born following rape or while the pregnant person is in an abusive or controlling relationship. While this differs from pressures placed on women to behave in particular ways to avoid risk during pregnancy, it similarly sends a message about who is, and is not, a 'good' mother. The cap creates an exception whereby people who receive welfare support are put in a position in which they are unable to access the choices about when to have children and how many children to have which are readily available to those with a stable income. And in a statement that echoes some of the more jarring language used by Haldane and his peers about curbing the reproduction of the working class, the policy explicitly invites women to 'think very carefully about whether they can afford to support additional children'.

The idea that some groups of women pose a special 'risk' to their fetuses informs the Pause project, too. Pause is a charity based in England that has received funding from the Department for Education and has twenty-eight practices in thirty-eight local authorities across the UK so far. The programme, which is targeted at women who are struggling with multiple challenges such as criminal records and homelessness, offers participants a tailored support package if they agree to use contraceptives for the eighteen-month duration of the programme. Pause works on a model of voluntary compliance: an agreement to use long-acting contraceptives (LARC) in exchange for support and rehabilitation. But given the lack of alternative broad-reaching support programmes for women who, as many of the members of the Pause project have, face issues ranging from domestic violence to homelessness and drug and alcohol use, the voluntary choice to use long-acting birth control occurs under circumstances where the opportunity for choice is limited. As critics have noted, without the option of other avenues for intensive support that do not require contraceptive use, this is a practice that raises serious concerns about reproductive rights and freedoms.

The programme harks back to other instances, in the United States in particular, beginning in the 1920s, of welfare support being granted only in exchange for sterilisation or agreement to use long-acting contraceptives. Pause requires people who have limited avenues for the forms of care it provides to cede control over their reproductive bodies. It reinforces an old hierarchy wherein wealthy and middle-class people with resources can access any form of birth control they choose, alongside the resources they need to promote their well-being, while those with limited resources of their own are forced to hand over autonomy in order to access support that a just society would provide them with anyway. This practice of birth-control use in exchange for care does not travel so far from Norman Haire's suggestion in his 1920s contribution to the *To-Day and To-Morrow* series that 'if either of the parents is unhealthy it may be necessary in the interest of the unborn child to prescribe avoidance of parenthood either for a time or forever'. The idea that the state might fairly decide that it is in 'the best interests' of some parents for their children not to be born into the world at all has remained insidiously persistent.

A hyper-focus on the safety of fetuses and the actions of pregnant people is not about protecting children. If that were the case, the same people who advocate treating women's wombs as a 'controllable environment' would also advocate against the state separating immigrant children from their parents. They would advocate against pregnant women and new mothers being incarcerated. Instead, this is an ideological approach that understands pregnant people's bodies as public property, and that exerts control over gendered and racialised people by exerting control over reproduction. While the tools for state-led projects of controlling reproduction have changed, purported concerns on the part of the state about irresponsible behaviour in pregnancy, and 'fitness' to parent, are part of the ongoing racist tradition of eugenics. Imagining that we might use artificial-womb technology to take over pregnancy where particular groups of people were believed to be 'dangers' to their future children is not a neutral project. And as we know, it isn't a new one.

What allows contemporary commentators in the 2020s to surmise that there might be an acceptable way to use artificial wombs to protect fetuses placed at 'risk' by the behaviour of their mothers? So long as eugenics is understood as something only undertaken by conservative or authoritarian states, rather than as a practice that has been engaged by governments and institutions across political divides, its legacy is able to carry on. Musings in the 2000s that ectogenesis might be 'safer' for fetuses than some pregnant people's bodies are often not coded as eugenics. The belief that we no longer live in a society where eugenic ideas are permitted in the mainstream enables such thought-experiments to occur with relatively little pushback. But we don't have to look much further than the British government's tacit acceptance of unnecessary deaths from Covid-19 to see the potent persistence of the hierarchisation of lives that circulated in the 1910s and 1920s. Jurisdictions with populist leaders, like Britain, the United States (during the first several waves) and

Brazil, were particularly notable for the absence of measures to protect those most vulnerable to the virus. Yet countries with liberal governments, like Canada, also saw the popularisation of rhetoric implying that because those who were dying were older people, the immunocompromised and those with underlying health conditions, they were somehow insignificant. As measures for managing the virus were lifted, political leaders and much of the media amplified the idea that it was 'only' the already susceptible who were in harm's way, and therefore it was time to 'live with Covid'.

The legacy of categorising people as being more or less worthy of life – and as 'good' or 'bad' candidates for reproducing – is not just a through-line from Ludovici to conservative bioethicists and scientific researchers. It also reaches from leftists who fought for reproductive rights and gender equality in the 1920s through to contemporary liberal and left-wing institutions, states and activists. Among the contemporaries of Haire, Haldane and Huxley who had been involved in both progressive campaigns to promote gender equality and in promoting eugenics were two of the leading figures in British and American feminist activism in the 1920s: Marie Stopes and Margaret Sanger. Stopes had opened the first family-planning clinic in London in 1921 and challenged the opposition of the Church, and of many in the medical profession, to provide contraceptive guidance and resources to women. She also, however, called for the sterilisation of those she deemed 'hopelessly rotten and racially diseased'[16] and opposed interracial marriage. In America, Margaret Sanger, the founder of the Planned Parenthood Federation, spoke about the importance of women having control over their own reproductive bodies, even as she endorsed the *Buck v. Bell* ruling.

In 2020 the Marie Stopes International officially changed its name to MSI Reproductive Choices, publicly commenting that Stopes's values were not aligned with the aims of the organisation. Planned Parenthood, in the same year, released a statement disavowing Sanger's eugenicist alignments as inherently ableist and racist, and simultaneously debunking false, exaggerated claims made about her beliefs by anti-abortion organisers.

Despite changes over the last several years, as scholar and social-justice advocate Dorothy Roberts argues, the modern pro-choice campaign, and the left overall, has not been fully purged of its eugenicist roots.[17] The idea that a forward-thinking position on reproductive life is one that simply grants people the right to end or prevent pregnancies fails to account for the fact that for hundreds of years Black women, Indigenous women, women of colour, people with disabilities and queer and trans people have also had to fight against sterilisation and the forced removal of their children.

In the 1990s a caucus of US-based Black feminists developed the reproductive justice framework at a pro-choice conference, initiating a grassroots movement that challenged the way the rhetoric of reproductive rights had long focused on the interests of white women. The founders of reproductive justice, and the activists, healthcare practitioners, lawyers and educators who continue to put this framework into practice, understood that a movement for reproductive freedom that is truly inclusive cannot take a sole focus on protecting a right to end or prevent a pregnancy. Instead, reproductive justice is an approach that recognises the equal importance of fighting for 'the right to have a child; the right not to have a child; and the right to parent the children we have, as well as to control our birthing options', as well as 'the necessary enabling conditions to realize these rights'.[18] Reproductive justice activists understand that each of these pursuits is inextricably linked with the pursuit of resources and of structural, systemic change. When you consider reproductive rights alongside racial, gendered, classed and environmental justice, when you understand these issues of social justice as inherently entangled, it is no longer possible to entertain the idea that state or institutional control over any person's reproduction is permissible. The reproductive justice movement's holistic approach to thinking about reproductive life demands an understanding of justice in which any categorisation of people as 'good' or 'bad' candidates to reproduce is a perpetuation of harm.

There is no acceptable or ethical way for a state or institution to prevent people from having children. Whether helmed by Nazis or leftist activists, this is a violation of human rights, a practice that takes people's reproductive autonomy away from them. Yet part of what has enabled these activities to persist is the failure – not only of Huxley, but of entire nations – to recognise that the problem is not just the danger of eugenics in the hands of Fascists. It is the very idea that human beings can be classified and placed in a hierarchy. It is the existence of any system by which people are categorised as being more or less worthy of life.

We may have travelled a long way since the days when J. B. S. Haldane stood in front of a weary but optimistic assembly of young scientists, academics and writers and pondered the way in which ectogenesis would be used in the future to create 'better' humans. Scientific research has progressed beyond many of his conjectures. But can we really say that we have built a world where we could trust that artificial wombs would not be used to coercive ends?

The writers who contributed to the *To-day and To-morrow* series did not all share Haldane's take on ectogenesis and eugenics. In Vera Brittain's *Halcyon, or the Future of Monogamy* an erudite Professor Minerva Huxterwin tells of the scientific developments leading up to the 2050s. Huxterwin describes an alternative lineage to that traced by Haldane. Universal ectogenesis became possible, she notes, but society ultimately rejected its widespread use: the kind of 'mothering' care that could be provided by both parents far outstripped what could be offered by the state and a machine. Care, she imagines, would be revealed to have a scientific and measurable value. While the 'ectogeneticians' would celebrate the way artificial wombs allowed them to propagate the most 'fit', their focus on genetic determinism would prove misguided. She suggests, in other words, that while people would try to use ectogenesis to perfect the 'science' of eugenics, they would soon find that a child's future depended not on their heredity, but on the love and affection provided by their parents. The story Huxterwin tells of artificial wombs is one in which this form of gestation remains available, but is used only when parents cannot conceive without it. Instead the future differs from Brittain's own, not primarily because infants are grown through ectogenesis, but because all parents are provided with sufficient resources to care for themselves and their children, and to manage their reproductive lives.

All else being equitable, families built on love reproduce a flourishing society. It is telling that this vision of the future remains as speculative now as it was then. As another *To-day and To-morrow* contributor, the scientist J. D. Bernal, put it, extrapolating on the realities of our present to explore the possible dangers that lie ahead is a way to 'see the effects of our own actions and their probable consequences'.[19] Considering the coercive practices for controlling reproduction that continue today, what are the dangers of how artificial wombs might be used in our future?

What does it say about us that it is easier to imagine a world in which babies are gestated outside human bodies than it is to imagine a world in which we have abolished eugenic policies and systems of classification? It should be a small and modest request to live in a reality where all people are provided with the resources they need and the space to freely decide for themselves whether, and when, to have children.

4

Mother Machine

When we consider the ongoing legacy of eugenics, there are good reasons to be attentive to the ways in which artificial wombs could be used to cause harm. It is important to remember, however, that the intended use of the technologies currently in development is to support extremely premature babies and pregnant people. Far from undermining people's control of their reproductive bodies, the researchers who are creating these platforms today aim to provide lifesaving assistance in the traumatic circumstances of preterm birth. The technology has, rightly, been celebrated as a potential game-changer for the health of neonates and pregnant people. But let's assume for a moment that artificial wombs will only be applied for this purpose: as a new form of reproductive care. To whom, exactly, will they be available? Today, in the absence of ectogenesis, access to existing forms of basic healthcare remains nearly as out of reach to many around the globe as it was when Vera Brittain penned *Halcyon* in the waning years of the Jazz Age.

The technologies currently in development are anticipated to be extremely costly and labour-intensive and to require a significant infrastructure to keep them running. In trials of the

EVE platform, researchers continuously monitored the growing lamb fetuses over twenty-four-hour shifts. If the final iteration of their artificial womb needs the same round-the-clock supervision, then it will probably necessitate carefully training an expert team. Researchers in the Netherlands, meanwhile, hope to produce a state-of-the-art prototype that will be trialled with eerily mammalian 3D-printed mannequins before it is ready for clinical use. To reach this goal, they will source parts from multiple biotechnology partners, potentially rendering their device very expensive indeed. In the United States each feature of the EXTEND project must be closely managed, including an artificial placenta that removes carbon dioxide and delivers nutrients to the fetus, a monitor that constantly tracks heart rate and blood flow, an adaptable catheter that functions as an umbilical cord, and the artificial womb itself. It also requires the production of approximately 300 gallons of carefully measured artificial amniotic fluid per day. Sourcing, assembling and maintaining any of these elements on their own would be challenging enough, but put all these parts together and building the artificial womb – and making sure that it runs as it should – will demand an immense amount of effort and oversight. As one of the lead researchers working on the EVE platform opined to a reporter for *Slate* in 2020, 'The practical reality is that this is not a fun discretionary birthing option, like a water bath. Assuming for a moment that we are going to get this to work, it will be eye-wateringly expensive and require an extraordinarily skilled team of people.'[1]

Why does any of this matter? The cost of the technology, the multiple components that are needed to apply it, the substantial infrastructure required to keep it stable and the necessity of expert supervision mean that artificial wombs are likely to be available only in already well-resourced environments. At the beginning of the pandemic there was much public discussion about how we would work towards health equality for all, at the end of the crisis. As of spring 2022 the injustice of how care resources are distributed remains as stark as ever. Even as wealthy nations declared a return to 'normal', most of the world still had not been vaccinated. Most of the world still could not access treatment if its inhabitants became infected with the virus.

Some of us get to move through our days knowing that we could stroll into a pharmacy and receive a life-preserving vaccine. And some of us get to experience our pregnancies knowing that we could be at a hospital within twenty minutes and receive respectful, thorough and expert attention. It has been all too easy, over the last few years, for much of society simply to look away from the excruciating extent of healthcare inequity. But these inequities are a damning indictment of the kind of world we have built, one that treads closer to Huxley's dystopia than to the utopia of collective care contemplated by Brittain.

It should matter to everyone that people in wealthy nations received three or even four doses of vaccine before those in low-income nations received one. And it should matter to everyone that some of the cruellest consequences of the inequitable distribution of crucial care resources fall on pregnant people and neonates. Artificial wombs could save the lives of extremely preterm infants who would otherwise die. A baby born at twenty-two or even twenty-one weeks – no bigger than a carrot – could grow and flourish, transforming the experience of extremely preterm birth. Further in the future, artificial wombs could benefit pregnant people who experience complications that endanger their health or lives in their late second or third trimester. Someone who today might be at serious risk and require close monitoring due to pre-eclampsia could perhaps be treated using an artificial womb instead. Yet this technology is being introduced into a context of extreme health inequity and, on its current path, is likely to be available only to a lucky few.

Approximately 15 million babies around the world are born preterm each year. Preterm births occur at the highest rates in India, China, Nigeria, Pakistan, Indonesia, the US, Bangladesh, the Philippines, the Democratic Republic of the Congo and Brazil. The World Health Organization reports that 94 per cent of maternal deaths globally (that is, deaths of pregnant people in the period before, during or after birth) occur in low- and lower–medium-income countries, with 86 per cent occurring in sub-Saharan Africa and southern Asia. Infants in these regions are ten times more likely to die in the first month of life than those in high-income countries. In low-income countries 90 per cent of babies born before twenty-eight weeks die in the first days of their lives, compared to fewer than 10 per cent in high-income countries. Extremely preterm infants, born at less than twenty-eight weeks, are precisely the target group for whom the partial artificial womb is intended and for whom it has been heralded as a revolutionary innovation.

In some of the world's wealthiest nations where healthcare resources are more readily available, outcomes for pregnant people, mothers and infants are marked by striking racialised inequity. In the US, Black pregnant women die at three to four times the rate of white pregnant women and are substantively more likely to have 'near-miss' experiences and to sustain injuries and health complications related to pregnancy and birth. Indigenous women are estimated to have a 4.5 times greater likelihood of dying from causes related to pregnancy and birth than white women in urban areas. Black, Native Hawaiian, American Indian and Alaska Native[2] infants in America are at an increased risk of being born premature, and more likely to die within their first year of life. In some states the gap in health outcomes between Black mothers and infants and their white counterparts is in fact widening. Between 2005 and 2014 Alaska Native and American Indian babies were the only infants for whom there was not a decrease in mortality.

If global and racialised disparities in rates of preterm deaths were a consequence of a lack of advanced neonatal technology, it might be most fair to ensure that partial artificial wombs were distributed accordingly. After clinical trials demonstrated that the technology was safe and effective, prioritising those most at risk might mean allocating artificial wombs to nations with the highest rates of deaths linked to preterm birth, for instance. If we know that rates of preterm birth are low in Britain and that even very premature infants born there have a good chance of survival, but preterm birth rates are high in India and these infants face a considerable possibility of death, couldn't we make the case that India should have the first access to artificial wombs?

However, the absence of advanced technology is not the only, or even the primary, challenge here. The WHO has noted, in fact, that in some low-resource settings where neonatal technologies are available, their use in the absence of the infrastructure they require has sometimes contributed to the problem of infant death. The application of technologies such as incubators that require a stable power flow, for example, in contexts where electricity cuts in and out, can put premature babies who might otherwise benefit from this treatment at risk. Even if artificial wombs were provided for free or at low cost, they will still require significant human labour and a substantial, predictable infrastructure: power supply, facilities in which to create artificial amniotic fluid, space to prepare each of the constituent parts. It wouldn't be enough to simply direct the artificial womb to clinics and hospitals without these resources. Where there is already a shortage of health personnel, a technology

that demands significant training to use it, and ongoing maintenance, would also mean redirecting practitioners away from
already pressing care duties.

In other words, inequity is never 'solved' by new and better
technologies alone. What is really at issue is the unjust distribution of basic resources, and not novel innovations. The WHO
estimates that approximately half of all babies in low-income
nations born below or at thirty-two weeks' gestation 'die due to a
lack of feasible, cost-effective care, such as warmth, breastfeeding
support, and basic care for infections and breathing difficulties.'[3]
Unsurprisingly, it is often the support that is available to the
pregnant person that can have the biggest impact on both the
likelihood of preterm birth happening to begin with and on subsequent outcomes for both babies and birthing parents.

While we know that there are scientific and ethical hurdles
to be overcome before artificial wombs are available in a
medical setting, let's imagine for a moment that clinical trials
occur within the five-to-ten-year timeframe predicted by the
EXTEND researchers. Let's then assume that they prove to be
just as effective as anticipated. By 2032 we might have a platform that facilitates the survival in good health of fetuses born
at twenty-one to twenty-three weeks' gestation. But what if there
is little improvement in the provision of antibiotics, steroids and
trained and culturally responsive health professionals in low-
income nations over the same period? If this scenario plays out,
artificial wombs will increase the already extreme inequity in
the experiences of pregnant people and babies between wealthy
nations and low-income ones. This is a bleak thought-experiment, but it is also an honest one.

Novel technology is not the solution to racialised health inequity for neonates and pregnant people, either. These disparities
are instead symptomatic of social problems – namely, structural

racism, discrimination in medical institutions and deep-rooted racialised inequity. In the United States racism is a determining factor in whether a person has access to sufficient medical insurance, and whether they live near a well-funded healthcare clinic. A 2019 study traced very literal segregation in NICUs, finding that Black babies were more likely to be concentrated in lower-quality NICUs than white, white Hispanic and Asian infants.[4] This is not just about the distribution of resources, however. In the US privileges such as education and financial stability impact upon health outcomes for white pregnant people and their babies, but Black women and their infants are put at risk during pregnancy by racism, regardless of their educational or economic status.

Black patients experiencing delays in or denials of care after reporting symptoms and pain constitutes a recurrent theme in research on racialised disparities in health outcomes for pregnant people and infants. The legacy of eugenics and unethical, racist human-subject research is deeply relevant here. Not so long before Martin Couney began his tours, J. Marion Sims, the inventor of the speculum, conducted painful and invasive surgeries on enslaved Black women without anaesthesia to develop techniques that he would later perform on sedated upper-class white women. As Harriet Washington documents in *Medical Apartheid*, he also conducted violent and fatal experiments on Black babies, blaming their deaths on their mothers and on the Black midwives who delivered them.[5]

The myth that Black women are less sensitive to physical suffering than white women – on which J. Marion Sims staked his experiments more than a century ago – still permeates care for mothers and infants. As Dána-Ain Davis argues in her book *Reproductive Injustice: Racism, Pregnancy, and Premature Birth*, this perception has been extended in some cases to premature babies. She cites the ongoing perpetuation of the unscientific claim that Black and Asian infants 'required shorter gestational lengths' than white babies. The danger of these kinds of ideas of course, is non-white infants being 'viewed as superhuman or in need of less care'.[6] Imagine, again, the scenario in which an effective artificial-womb platform is available in ten years' time. Introduced within a culture and institutional structure where racism persists, where Black prematurely born babies do not seem to be given the same priority for treatment as white babies, the technology could improve already better overall outcomes for white pregnant people and their babies, while leaving disproportionate rates of death and complications for Black infants and their mothers untouched.

The United States is often cited as an extreme example of racialised health inequity, and it is unique among many high-income countries for its appallingly high rates of maternal and neonatal mortality overall. Yet the persistence of racism as a health risk is evident even in countries with much lower incidences of preterm birth and maternal and neonatal mortality. As the researchers leading the EVE study recently noted, in Australia, where their work is partially based, 'the rate of preterm birth in Aboriginal and Torres Strait Islander populations is nearly double that of the non-Indigenous population'.[7] In Britain, where the NHS provides coverage for all pregnant people, Black women die of pregnancy-related complications at four times the rate of white women, are more likely to suffer from avoidable health complications associated with pregnancy and more likely to have preterm births. Asian and mixed-ethnicity women, too, die at substantively higher rates and are more likely to have a preterm birth than are white women.

The partial artificial wombs currently being developed are lauded for their potential to redress some of the very conditions from which minoritised women and infants are more likely to suffer. Babies born preterm, disproportionately, to Black women in America and the UK, and to Pacific and Torres Strait Islanders in Australia, might be gestated safely to the twenty-eight-week point when their chances of survival in good health could drastically increase. But just like inequity in health outcomes for pregnant people and babies between low- and high-income countries, this is not a problem that technological innovation, in and of itself, could ever address. An artificial womb is a means of intervening after an extremely preterm birth has already occurred, to reduce the likelihood of death and health complications. It does not, of course, do anything to prevent disparities in the rates of preterm birth to begin with.

Dr Joia Crear-Perry, ob-gyn and founder and president of the US-based National Birth Equity Collaborative observed in an article for *The Root* that there is a regrettable tendency to focus on technological solutions rather than on social strategies to fix social problems. As a consequence, she noted, 'investment in biotechnology and not people leads to the improvements of things and not human beings.'[8] Both artificial wombs and the means by which they may be allocated drift towards an emphasis on novel, high-tech strategies for addressing preterm birth, rather than social ones. In a 2017 paper discussing the success of their first animal trials, the EXTEND team speculated that 'future developments may allow better prediction of those infants who are destined for extreme premature delivery and may allow genetic prediction of infants who are most at risk for mortality and morbidity if born premature'.[9] They suggest that these possible developments could support decisions about which babies should receive artificial-womb care. This is no doubt a logical way to proceed, from a scientific perspective. Theoretically, at a future hospital in the US with, say, one artificial womb in its possession, genetic prediction might be used to establish that, among three pregnant people likely to give birth prematurely, one of the three preterm babies is highly likely to die or suffer adverse outcomes. The other two might be assessed to have a good chance of survival and a low risk of complications using other forms of care. The artificial-womb treatment, then, is reserved for the infant predicted to do most poorly without it.

But would this approach leave social determinants of health entirely out of the equation? We don't need genetic prediction to assess who is most at risk of preterm birth, and whose infants are most at risk of ill health or morbidity if born prematurely. Uneven distribution of resources and the impact of structural racism act as clear social indicators of which preterm infants are most in danger in America. While artificial-womb technology couldn't negate the injustices that produced these disparities in the first place, there is already a substantive body of evidence in the US that demonstrates a causal relationship between social factors and deaths and complications associated with preterm birth. If the goal is to determine who should have access to a limited, lifesaving resource, why should genetic data be prioritised over this evidence in making decisions about distribution?

An argument could be made that focusing on genetic assessments, as opposed to social factors, would reduce the possibility of bias. If a team of healthcare practitioners and ethicists had to allocate artificial wombs based on social determinants of health, these might be subjective assessments of risk, in contrast to assessments based on scientific data. As we know from thinking about the legacy of eugenics and human-subject research, however, scientific and technological intervention do not occur in a vacuum. Existing discrimination is often mapped onto the application of high-tech tools that are proclaimed to be neutral, both in the way these technologies are created and in the way they are applied. Take, for instance, face and voice recognition. The use of datasets drawn primarily from white men to create these technologies in the first place means that their error rate for this group is extremely low, while their error rate for dark-skinned women is over 30 per cent. And the use of these technologies in contexts where racism already enables significant harm, such as in policing, sees them applied as new tools to continue an old practice of surveilling and targeting Black people and people of colour.

What steps would be taken to try to ensure that genetic prediction to determine the risks associated with preterm birth was in fact a neutral activity? Who would be responsible for assessing the data that was collected? How likely is it that someone making a risk assessment would look at the data alone? Might their identity, experiences and social context factor into their recommendations? Returning to our hospital scenario in which only one artificial womb is available, imagine a situation in which five extremely preterm births occurred over several days. Each twenty-two-week infant is determined to have a good chance of survival, with significantly improved health outcomes if placed within the artificial womb. Could a truly impartial determination about how the technology should be applied be reached, using solely genetic data? In the US, financial considerations might also play a role in these decisions. Would care go to the patient whose parents could pay for it or the patient assessed to be most likely to die without this treatment?

Preterm birth and health issues that arise for pregnant people in the latter stages of gestation are tangible harms that can have devastating consequences for people's lives. We can celebrate the creation of tools that might help redress these harms, and the ingenuity of a method that could quite literally enable a baby to develop as though she remained in the womb. But we must be realistic about what these technologies can and cannot do on their current trajectory. How do we chart a different path forward – one where artificial wombs do not increase existing injustice – when both global and racialised health inequity are so deeply engrained?

How do we reach towards a future in which everyone is granted sufficient resources during and after pregnancy? As we saw in Chapter 3, eugenic ideas that some people's lives are more valuable than others have become entrenched in medicine and law. Hospitals and clinics across wealthy nations in which Black and Indigenous women and their babies are placed at risk by racism and discrimination are shaped by these legacies. In North America regulatory frameworks that continue to limit the activities of midwives are a relic of intentional efforts by a white, male-dominated medical establishment to take reproductive care out of communities and exert control over traditional carers and Black and Indigenous midwives. At the level of government policy, and all the way down to how students in medical schools are taught, failures to provide equitable care for pregnant people and infants persist within a system that was built to value white mothers and babies over others.

While a form of external gestation that empowers pregnant people rather than undermining them remains speculative for now, very real work is being done in the present to confront health inequity and improve experiences of pregnancy, birth and parenthood. Protecting reproductive healthcare as a fundamental human right in practice is a precondition for building a future in which artificial wombs might actually benefit all pregnant people and neonates.

In wealthy nations like the US and the UK, grassroots organisers, healthcare practitioners and policymakers have been taking substantial steps to confront racialised health inequity and protect reproductive care as a human right. In the United States, organisations led by a reproductive justice framework have used a human-rights approach to 'treat [...] reproductive health services as akin to the resources all human beings are entitled to – such as health care, education, housing, and food'.[10] In their collection *Radical Reproductive Justice*, Loretta Ross, Lynn Roberts, Erika Derkas, Whitney Peoples and Pamela Bridgewater Toure emphasise that alongside fighting to protect pregnant people's self-determination, protecting reproductive care as a human right means fighting for 'the obligation of government and society to ensure that the conditions are suitable for implementing one's decisions'.[11] In other words, it isn't good enough simply to focus on securing reproductive choice. People need access to free, safe, local reproductive healthcare throughout the course of their lives, including contraception, abortion, prenatal support and respectful birth and post-partum care. And societies need to guarantee this access and actively protect it.

Among numerous organising efforts in the US, the National Birth Equity Collaborative's Birth Equity Agenda, a comprehensive blueprint for change, offers one example of what it might look like to protect reproductive health as a fundamental right. The Birth Equity Collaborative advocates for reproductive health and autonomy to be granted proactive protections at all levels of government. Among other recommendations, it proposes the creation of a White House office that would engage a human-rights and racial-equity approach to addressing all 'barriers to full reproductive autonomy, such as access to health care, including contraception, maternal and infant health, quality, affordable childcare, and comprehensive paid family leave'.[12] At the level of policy, these kinds of practices speak to the social – not technological – innovations that could shift the needle on perinatal and neonatal health inequity.

Protecting the health of infants and pregnant people as a human right and ending inequity in care is a global issue. In the 2020s one of the United Nation's sustainable development goals includes reducing stillbirth and maternal and infant mortality and morbidity in the low- and middle-income countries where most of these events are concentrated. These improvements continue to require significant international resources. Given the reluctance of wealthy countries to share essential medicines like the vaccines that might have brought a swift end to the pandemic, how likely is it that these same countries would be willing to share the means of improving health for pregnant people and infants?

The WHO reports that in the eighty-eight nations where 95 per cent of the world's stillbirths and newborn deaths occur there are significant shortages of healthcare professionals. Although these regions are home to 74 per cent of the world's population, they are host to only 46 per cent of the world's doctors and nurses. A joint study by the WHO, the United Nations Population Fund and the International Confederation of Midwives published in *The Lancet* in 2020 found that where culturally appropriate, high-quality midwifery care was widely available, two-thirds of all maternal and neonatal deaths and stillbirths could be prevented each year by 2035.[13] This would also require supporting midwives with training and sufficient pay, and providing access to supplies and sanitation. Protecting reproductive care and care for neonates as fundamental human rights in practice means backing communities and grassroots movements that have long been working towards these ends, with resources, funding and attention.

Until the kinds of social changes that grassroots initiatives, progressive policymakers and healthcare practitioners are driving forward globally are implemented, the possibility that artificial wombs could truly be groundbreaking for preterm babies and pregnant people in general will remain out of reach. What would the world have to look like before this future could be realised? Any person, of any gender, any race, class or ethnicity, wherever they lived, would need to be able to access safe, culturally appropriate and thorough reproductive care. They would be able to choose the support of midwives, obstetricians and doulas during their birth, for free, and with the knowledge that their care providers would respect them and protect their safety and that of their babies. And they would be able to give birth knowing that their babies would have access to the resources they needed in order to thrive.

I often find myself questioning whether funding expensive biotechnologies that will remain inaccessible to most can ever be justified, when we distribute and apply existing care supplies in such an inequitable way. Wouldn't the just path forward be to turn away from costly innovations like artificial wombs and towards efforts to tackle inequity?

The challenge here is that artificial-womb research has already received substantive funding and support. Partial artificial wombs are not hypothetical or slated to arrive in the distant future. In 2019 the Dutch research team alone received €2.9 million towards the initial phase of their project: producing a functional prototype. It is not a question of whether it is better to grant resources towards this work or towards, say, initiatives to prevent preterm birth in the first place. The reality is that this research has already been funded and is already happening. We know that the solution to health inequity is the pursuit of social change, not technological progress. In the meantime, while artificial-womb studies are ongoing, we might also ask whether issues of access and equity could be considered in ways that help mitigate the technology worsening existing injustices.

In Chapter 2 we explored the extremely challenging path towards clinical trials with humans. When researchers are developing novel means of treating preterm babies, accessibility, adaptability to different environments and affordability – all of which might affect a technology's social impact – are rarely front and centre. But what if these issues were key features in the artificial womb's design? And could public consultations that centred those who are most marginalised in reproductive health-care allow for a rethinking of how the technology is developed, distributed and used?

When asked about the financial implications of external gestation in 2017, a researcher for the US-based group that had just announced the bio-bag commented that it was 'too soon to speculate', adding that any expenditure would be offset by reducing the expenses related to managing the health complications of preterm birth. Taking account of everything from initial hospital stays to ongoing treatment for lung disease, intestinal disorders and other long-term health issues caused by preterm birth, the cost of caring for all prematurely born infants in the United States alone is estimated at a staggering $26 billion per year. In the US many of the costs of neonatal care units are out-of-pocket services. Today a stay in an American NICU can run to $3,500 a day, with a long stay potentially exceeding $1 million. There are many people without medical insurance living in the US, for whom these payments are already prohibitive. There are others with private insurance for whom a NICU stay could mean running into debt or no longer being able to cover care for the rest of their family. If the cost of gestating infants for four weeks in a partial artificial womb fell to parents, then its 'eye-watering' expense would put it out of reach of most people.

It is often long after a technology is introduced that more accessible iterations are created. Existing patented incubators that are commonly available in high-income nations continue to run as high as $37,000–50,000 USD. But incubator parts can be difficult to source, and the technology can be challenging to use and maintain. As a consequence, even when conventional incubators are made available outside state-of-the-art NICUs, they are not always practical or functional. Over the last twenty years groups of researchers have worked to create iterations of the incubator that are both more financially viable and more adaptable. A medical technology start-up in Britain, for instance, has developed a low-cost, easy-maintenance, inflatable model dubbed 'mOm'. Embrace Innovations has produced a sleeping bag-like warmer that costs around $200 USD and is designed for the express purpose of being able to function in almost any setting. Because it does not require electricity to warm it and is easy to use, it can be lifesaving in contexts where mothers may be far from hospital care after delivering a preterm baby. These and other similar innovations speak to the potential for neonatal technologies that, while not at all fixing the problem of inequity in care, could help to address the consequences of that inequity.

Is it possible that artificial wombs could be created and implemented in such a way that beyond reducing the financial expense of preterm birth in countries like the United States, they could be used outside wealthy Western hospitals? Could focus during the development process on the overhead for materials used to make the artificial womb, as well as whether it could be easily engaged in low-resource settings, alter its impact for the better? What if researchers made it a priority to create a prototype that required minimal training and functioned safely in spaces with limited infrastructure for neonatal care?

There are equally important questions to be asked about the possible benefits of consulting with a broad range of people who are directly affected by health inequity for pregnant and birthing people and neonates, and who work in areas that are shaped by these issues. Each of the groups currently developing a prototype has expressed the intention to engage in discussions outside their labs, aimed at thinking through some of the implications of their research. The EXTEND team, which cited a three- to five-year timeline for human trials as of 2017, is working with bioethicists to explore an ethical framework for the project and the clinical contexts in which it could be used. The co-leads of the EVE platform have also noted that their work will raise significant legal and ethical questions. And in the Netherlands researchers have already begun to engage the public in discussions of the technology's impact and the possible ethical challenges it may pose. Who is considered a stakeholder in consultations about how the artificial womb is implemented might also influence the technology's social repercussions.

A researcher who works in neonatology may have different ideas about the possible challenges of using an artificial womb in a clinical setting than, say, an obstetrician or a neonatal intensive-care nurse might. Midwives, ob-gyns and doulas might raise questions about how artificial wombs could impact upon pregnant and birthing people's experiences that would not be considered by a neonatologist. A healthcare practitioner who worked with infants in a setting with few resources could offer knowledge about what might make an artificial womb functional for their patients that might not be reflected on by an academic researcher at a well-funded American clinic. Providers in these settings might beneficially inform the development of the technology – perhaps allowing for features that could better support its use in environments with, for instance, unstable access to electricity. Grassroots organisations that have been fighting to reduce preterm birth in low-income nations, and Black feminist-led groups pursuing health justice for infants and pregnant people in the US and UK, might have instructive insights into how to reduce the possibility of this technology worsening existing health inequity. What questions might these groups have about the way clinical trials will be conducted, or how access decisions might be made?

Finally, pregnant people and parents whose babies have spent time in NICUs would likely have different concerns than the technology's developers would. Prioritising people who have been most marginalised in reproductive health, including Black and Indigenous women and LGBTQ+ people, in engagement about the uses of an artificial womb might lead to informative discussions about the technology's application and limitations.

The artificial wombs currently in development, and future projects on external gestation, will be shaped by whether inequity and injustice among pregnant people and neonates are considered relevant matters. 'Access' has never been solely about the cost of a health technology, or where it is made available. It is also about who a technology is designed for, and who gets to inform that design.

These are issues that, realistically, might never be considered during the projects that are currently ongoing. Public consultation processes are significant undertakings. And researchers already face substantive obstacles to progressing their projects to application in a clinical setting. It is perhaps understandable, then, that their primary focus would be on ensuring that the technology works – and works well – rather than on whether it will be accessible or will exacerbate existing inequity. But the possibility of other research towards external gestation, research that hasn't yet begun, remains open. The technologies in progress are unlikely to be the only kinds of artificial wombs that will ever be developed. Asking questions about how these technologies should or should not be used, who should be consulted in their composition and who they will be created for pushes us to think about the consequences of introducing artificial wombs in an unequal society. It also enables us, though, to imagine how things might be different.

Could ectogenesis be created with justice in mind from the start? What would be different? Can we hope for the world to be a much more equitable place by the time this technology is introduced – one where reproductive health is truly protected as a human right? If we want artificial wombs to actually benefit all pregnant people and neonates in practice, we first need to confront health inequity and to secure reproductive care for all.

In an unjust world, no technology is a miracle in and of itself.

5

The Abortion Solution

I have a vivid memory of the first time I encountered the word 'ectogenesis'. I was rifling through various articles – reproductive technologies, rights and justice – trying to narrow my research topic in time for my first meeting with my potential PhD supervisors. I was especially curious about *in vitro* fertilisation, egg-freezing and any number of compelling areas where the questions that come up in law require a certain amount of speculative thinking. I stumbled across an article from the early 2000s in which the author mentioned in passing that among all possible future reproductive technologies, ectogenesis would pose especially novel challenges for the law. Who would be the legal parent(s) of a baby grown in an artificial womb? Could this raise interesting possibilities for communal child-rearing? What was the danger that an ectogenetic baby would wind up as the property of the state? Who would have access to the technology, and under what circumstances would they be able to use it?

My immediate thought, before I gave it a great deal of consideration, was how incredible this technology could be for women. I knew as many people who hated everything about being pregnant as I knew people who relished it. Most people's experiences fell somewhere in between. But for friends who endured extreme nausea throughout the months that they carried their babies, who developed gestational diabetes and had painful and complex birthing processes, pregnancy took an extreme physical toll. And then there were the many conversations I had had with friends who very much wanted children and planned to have an equal parenting relationship with their partners, but were frustrated by the social and physical ways that pregnancy seemed to dictate the terms of their participation at home and in the workplace. How could they divide things 'equally' if one partner could get parental leave for the duration of their pregnancy and the first months of the baby's life, but the other struggled to get more than two weeks?

And this – the most basic of impulses – is what struck me when I first read that word 'ectogenesis': what if people could gestate using an artificial womb instead of carrying their pregnancy? What if someone who was diagnosed with pre-eclampsia could elect to transfer their baby to an artificial womb, protecting their health while enabling a wanted baby to continue to gestate? And what if any person of any gender could choose to be the caretaker for a baby gestated in this way? I would quickly come to understand that each of these ideas presents a tangled web of social, political and legal questions. And that while some of these possibilities might come to pass if the technology were in the hands of feminist, anti-racist researchers, in the context of our contemporary society each could serve only to worsen existing injustices. However, in that first interaction with the idea of ectogenesis, my impulse was to think of all the possible ways that this technology might be mobilised to benefit pregnant and birthing people.

Not so long after I landed on my PhD topic, in the last months of 2016, embryologists successfully grew embryos in culture up to the legal limit of fourteen days. And then, in the first months of 2017, the EXTEND researchers announced their own breakthrough. Animal trials of a partial artificial womb had been a resounding success for the first time. Suddenly that word – ectogenesis – was everywhere. But despite the sci-fi connotations it conjures up, despite all the ways it could benefit pregnant people, and despite the questions it stirs that had drawn my attention to it as a research subject, one topic came to dominate in both scholarly and popular takes on the technology. Namely, the idea that artificial wombs will 'solve' the abortion 'debate'. From 2016 until 2022 I would continue to encounter academic and mainstream forums that had published some version of this claim, epitomised by a 2019 *New York Times* feature by the American humanist activist and journalist Zoltan Istvan, called 'The Abortion Debate Is Stuck: Are Artificial Wombs the Answer?' Much like other articles of its genre, the piece posits that artificial wombs will 'resolve' the long, fraught battle over abortion. Feminists would be satisfied because the fetus could be removed from the pregnant person's body, thus protecting their bodily autonomy; and anti-abortion advocates would be satisfied because the fetus would be able to live. Istvan, in fact, would later announce that he planned to run in the 2020 presidential elections as an independent candidate, promoting a promise to fund artificial wombs and solve the 'debate' once and for all. Despite his apparent belief that his proposal was a novel one, he is just one of many who have proposed that artificial wombs could be a magic bullet for the 'problem' of abortion.

As I would learn back in 2016, this idea that artificial wombs could allow women to either 'foeticidally womb empty' or 'non-foeticidally womb empty',[1] and the law would compel them to do the latter, has been floating around since the 1970s. Bro-bioethicists, whose bodies are not implicated in this form of speculation, often present this idea as though it is simply neutral food for thought. But if you have a uterus there is nothing neutral about an unwanted pregnancy, or the possibility that a fetus could be extracted, alive, from your body and transplanted into an artificial womb against your will. Unfortunately, in many parts of the world abortion continues to be treated as a polarised political debate rather than as essential healthcare, and access to reproductive rights remains unstable at best.

What does the popularity of the claim that artificial wombs could 'solve' abortion tell us about how this procedure is socially perceived? And could artificial wombs actually pose a threat to abortion rights? Attitudes towards abortion, and the way these rights are protected in law, vary from nation to nation, but even in countries where abortion is permitted, it is often heavily regulated and subject to significant restrictions that do not apply to other medical procedures. It is no coincidence that much of the most vociferous scholarship and media coverage of the artificial womb as a 'solution' or challenge to abortion comes out of the US, where abortion rights were long articulated in balance with a fetus's potential for life, and where a vocal anti-abortion contingent holds political sway. When I turned over the final draft of this book at 39 weeks pregnant, I had written in anticipation of the Supreme Court striking down the right to abortion carved out in the landmark *Roe v. Wade* case. With my months-old baby in my arms in June 2022, I read Justice Samuel Alito's description of *Roe* as 'egregiously wrong' with horror. The fall of *Roe v. Wade*, however, was not sudden: it was the final piece of a slow deterioration that began in the immediate aftermath of the Supreme Court's 1973 decision. It was precisely in response to the way the Supreme Court had ruled on reproductive rights that the idea of artificial wombs as an end to abortion first appeared in legal scholarship.

The arrival of the Roe case in the Supreme Court in the early years of the 1970s was part of a carefully calculated plan executed by pro-choice lawyers and activists. *Roe v. Wade* is a prime example of strategic litigation. Lawyers and activists in the movement for reproductive rights had been watching for litigants who were sufficiently impacted by the existing law that they might have the standing to come before the court. If they could find such a case, and if it made its way to the Supreme Court, they could argue before the justices that laws criminalising abortion were unjust, and they might be able to get these laws overturned. It was Jane Roe's case that ultimately presented the opportunity for her pro-choice legal team to levy their arguments, in the hope that laws criminalising abortion would be ruled an unconstitutional violation of women's rights. The result was a significant victory: the Supreme Court agreed that forcing a person to remain pregnant against her will undermined her right to privacy.

Establishing precedent is essentially a matter of reminding a court: you ruled in this way on a similar matter, and so it would be unfair if you made a different ruling here. In Jane Roe's case, her lawyers argued that a right to abortion should be understood in a similar way to other personal and family matters that the court had previously ruled could be protected under a privacy right. The Due Process clause of the Fourteenth Amendment of the United States Constitution reads that 'no state shall … deprive any person of life, liberty, or property, without due process of law'; and in other cases involving family matters and people's rights to refuse unwanted medical treatment, the Supreme Court had ruled that the wording of this part of the Fourteenth Amendment could reasonably be interpreted to mean that people had a protected right to privacy on certain issues. Roe's legal team successfully argued that abortion should be considered one of the areas in which people should have a protected right to make a private decision about their bodies – a right to 'choose' whether they did, or did not, want to continue a pregnancy, without interference by the state.

Roe established a legal right to end a pregnancy, and in the United States it marked an end to the era of women dying and being injured by botched underground abortions. It also offered hope that the stigma and shame that were intensified by the clandestine nature of seeking an illegal procedure might be lifted. But the legacy of the landmark case is complicated. The Supreme Court justices who ruled in *Roe* were careful to establish that this was not a ruling for abortion on request. There were to be substantial limits. They made it clear that the state had two competing interests: protecting the privacy rights of the pregnant person and protecting the life of the fetus. To manage this balance, they came up with a loose framework based on the stages of gestation. Early in pregnancy, for roughly the first trimester when the fetus had no capacity for independent life, the pregnant person's rights would take precedence over the life of the fetus. From approximately the beginning of the second trimester, states could regulate abortion more rigorously, so long as those regulations were 'reasonably related to maternal health'. And finally in the third trimester, from about 26 weeks when the fetus was considered 'viable' – or understood to have a chance of survival outside the womb – 'the State in promoting its interest in the potentiality of human life may, if it chooses, regulate, and even proscribe, abortion', unless it was necessary for the pregnant person's health or life.

Roe v. Wade was not a straightforward, unequivocal ruling in support of reproductive choice. It did not legalise abortion, nor did it affirm it as something akin to other forms of healthcare. It did not establish that states had a duty to provide abortion services. Instead, *Roe v. Wade* affirmed that: a) people had a right to choose abortion, but not necessarily a right to be provided with access; and b) the right to abortion was always in balance with a state's right to protect the *potential* life of the fetus. Seeking protection for abortion by fighting for its recognition under a privacy right was a strategic choice. But it also meant that in the immediate aftermath of *Roe*, anti-abortion state legislatures and campaigners at the federal level were able to move quickly to make getting an abortion in practice as difficult as possible. While a privacy right in theory protects people from the state dictating what they can and cannot do, it does not require that the state grant them resources. It does not require that the state, for instance, provide them with funds with which to pay for abortion, or require the state to ensure that everyone has an abortion provider in their community or the ability to travel to a provider if necessary. So while for those with resources (both financial and social) *Roe* ensured that states could no longer interfere in their choice to get an abortion, for those without resources the right to 'choose' often meant little in practice.

A key example of the limitations of *Roe*'s protections to abortion came in the form of the Hyde Amendment, which was passed by Congress in 1977 after a crusade by the vocal anti-abortion Republican Henry Hyde. The amendment, which has existed in various forms ever since and was made permanent in 2016, bans federal funds for abortion, except where a pregnant person's life is at risk or in cases of pregnancy resulting from incest or rape. In some states (like ID, KY, MO, ND, OK) as of early spring 2022, insurance to cover abortion was limited in private plans, and in others (like CO, IL, KY, MA, MS, NE, DE, OH, PA, RI, SC and VA) insurance coverage for abortion was restricted. People otherwise covered by Medicaid, federal employees and their dependants, military personnel and dependants, Indigenous peoples otherwise covered by the Indian Health Services Act and incarcerated women could not access the service unless they were able to pay privately.[2]

As Henry Hyde well knew when he fought for the Hyde Amendment in the 1970s, his proposal played on the limitations of the privacy right to protect abortion and appealed to the balance between the state's interest in the woman's privacy and the potential life of the fetus that had been established in *Roe*. Women would still be 'allowed' to choose abortion, so their privacy right was upheld, but they could not use federal funds to enable this choice. Of course this had precisely the effect that Hyde intended – that is, it made legally sanctioned abortion access nearly impossible for people reliant on federal funds, disproportionately Black and Indigenous women, low-income women and women of colour. Hyde himself commented, 'I would certainly like to prevent, if I could legally, anybody having an abortion: a rich woman, a middle-class woman, or a poor woman. Unfortunately, the only vehicle available is the … Medicaid bill.'

In the decades after the *Roe* ruling, public and political support for abortion rights varied widely across American states, and across partisan lines. States with anti-abortion legislatures – that is, those with strong political and social motivation against protecting abortion (such as Georgia, Mississippi, Texas and Alabama) – continued to pass measures to make abortions very difficult, if not impossible, to access. Targeted Regulation of Abortion Providers (TRAP) laws, for instance, were measures passed by anti-abortion state legislatures under the guise of patient safety to make it increasingly harder for providers to offer abortions. These laws set requirements (such as minimum measurements for the size of rooms in which abortions are provided) that had nothing to do with protecting pregnant people and everything to do with curbing abortion.

Anti-abortion activists and legislatures continued to find new ways of inhibiting access to terminations. In 2021 Texas Bill SB8 came into force. The law banned all abortion after six weeks and was carefully crafted with the express intention of evading lawsuits that might see it struck down as unconstitutional before it took effect. To put it simply, SB8 was designed so that it was not enforced by the state or state officials. If it were, this would have allowed plaintiffs recourse to sue the state officials responsible for enforcing the law on the grounds that it was unconstitutional before it was ever actually enforced. Instead SB8 enabled private citizens to bring lawsuits against people who perform[ed] abortions after six weeks and against anyone who 'knowingly engage[d] in conduct that aid[ed] or abett[ed] the performance or inducement of an abortion'. Those bringing these lawsuits could claim a minimum of $10,000. Essentially, without having the state ban abortion outright, which would have been in violation of *Roe*'s protections, SB8 worked to make every person in Texas an informant on anyone from service providers to someone driving their friend to get a termination after six weeks. A 60 per cent drop in abortions in Texas was recorded in the month after the law came into force, with a subsequent increase in abortions performed in neighbouring states, reflecting that those with the means to do so had begun travelling out of the region for procedures no longer available closer to home. In June 2022, the Supreme Court's resolution in *Dobbs v. Jackson Women's Health Organization* saw the highest court in the country rule that there is no constitutional right to abortion in America, and that the protections eked out by *Roe* and the case law to follow should never have been granted. Following the ruling, the Guttmacher Institute (a reproductive policy research centre) anticipated that twenty-six American states would move to ban abortion entirely, with thirteen enacting 'trigger laws', bans designed to go into effect as soon as the court released its decision.

While some commentators have suggested that artificial wombs will pose a new challenge to abortion rights in the United States, when you consider the limitations of the *Roe* ruling, the subsequent attacks on reproductive rights in America, and the fall of legal protections to abortion, it is more accurate to say that these rights have been under duress for decades. But, as those bioethicists who have pondered artificial wombs as the 'end' of abortion well know, in places where abortion rights continue to be framed as a balance between a pregnant person's interests and a fetus's capacity for life, the technology is likely to be another tool deployed to attempt to further limit legally permitted abortions. Put simply, in any jurisdiction where protection to abortion in law is tenuous, and where vocal anti-abortionists hold political sway, artificial wombs could pose a threat.

How would abortion law have to change before we could be sure that this technology was not used to further draw into question people's ability to end their pregnancies, on their own terms? In a stuffy conference room during a London heatwave in summer 2018 I once again watched a bioethicist ask, 'If artificial wombs existed, would women be permitted to have abortions?' He went on to explain that, from an ethical perspective, abortion is permissible because it would be unjust to force someone to carry a pregnancy against their wishes. But if artificial wombs were introduced, a woman could have a fetus removed from her body into an artificial womb without causing its death, and she might be morally obligated to do so, rather than terminating it. He commented that although under contemporary circumstances, it would be unethical for a man to try to stop a woman from terminating a pregnancy, with artificial wombs each partner would have an equal claim to the fetus.

In the world beyond that conference room, the fight for free, legal, accessible abortion was ongoing. Pregnant people across the United States were facing fresh attacks on their reproductive autonomy as state legislatures emboldened by Trump's governance continued to chip away at access to abortion. In the Republic of Ireland the 'Repeal the Eighth' campaign celebrated a milestone in the long fight for pregnant people's self-determination, after the winning vote to repeal archaic legislation that held the life of a fetus above that of a pregnant person. In Argentina campaigners rallying for an end to their own restrictive abortion laws had been dealt a significant blow when the Senate narrowly voted against a bill to strike them down. Conference attendees heard hypothetical musings about a future technology that might mark the end of abortion rights. Meanwhile around the world – and as close to that room as an hour-long flight to Ireland – existing restrictions to abortion, in the absence of artificial wombs, continued to leave people at risk of death, injury and criminalisation. In 2012 thirty-one-year-old Savita Halappanavar, who was denied a procedure to complete the termination of a miscarriage, died of sepsis. In August 2018 a young Argentinian woman known only as Elizabeth died of an attempted self-abortion, not long after the Senate voted against revising the country's restrictions.

Where abortion is illegal, or nearly so, pregnant people have died – either by suicide or by being refused medical treatment on the grounds that it might inadvertently kill the fetus. These deaths are preventable. Part of the tragedy is that since the 1980s abortion procedures have been safer than ever. Throughout the 1960s and 1970s pro-choice advocates in nations like the United States and England shared stories of women's deaths by botched, back-alley abortions to garner support for lifting restrictions. But in the 1980s medication abortion, which involves taking or inserting two pills (mifepristone and misoprostol), made termination safer and easier than ever. Today, in the form of these pills we have the means for people to safely use medication in their own homes and end their pregnancies. And yet still, for the crime of seeking autonomy over their own bodies, women die, are harmed and face criminal charges.

As of 2022, in twenty-four countries abortion is not permitted under any circumstances. In forty-two countries it is permitted only to save the life of the person seeking the procedure. In fifty-six nations it is permitted to preserve life or health. In fourteen countries, which are generally perceived to have liberal abortion laws, including Great Britain, abortion is permitted on broad social and economic grounds. Yet in many of these nations, like Great Britain, abortion is still governed under criminal law. In seventy-two countries abortion is available on request – that is, it is not necessary for a person to meet certain conditions for approval to be granted. Many of these countries retain gestational limits after which abortion is more heavily restricted or cannot be accessed.

Like the controversy that surrounds embryo research, whether abortion is criminalised or permitted within a given jurisdiction depends on cultural factors, including the dominant religion within the region and the political influence of these groups. In England leading membership societies for medical practitioners (the British Medical Association and the Royal College of Obstetricians and Gynaecologists among them) have been supportive of protecting abortion on demand and removing criminal sanctions entirely. In Canada abortion is decriminalised throughout pregnancy (though there are gestational limits in some regions) and there is strong protection to abortion in law. But even in these jurisdictions, where abortion is broadly accepted as a settled issue, political support for the procedure often falls along partisan lines. In 2020 Conservative MP Fiona Bruce (Chair of the All-Party Parliamentary Pro-life Group) proposed a bill that would have created new limits on abortion across the UK. The bill failed to progress beyond a first reading in the House of Commons. In Canada in 2020 Dr Leslyn Lewis campaigned for leadership within the Conservative Party, partially on the promise of introducing new bans on some abortion procedures. In both cases, they drew on anti-abortion myths to present themselves as champions of both women's rights and the rights of the fetus. While each effort was unsuccessful, they received support from fellow conservatives. I returned to this chapter in early summer 2022, after the fall of *Roe*. I write this knowing that because of the way reproductive rights have been allowed to be intertwined with politics, the landscape of legal protections to abortion will have shifted yet again by the time this book goes to print.

Abortion wasn't always an object of bipartisan political debate, though. And the reasons for its early criminalisation help to explain why conservatives cling to dialling back reproductive rights, even in nations where access to the procedure has broad support among both doctors and the public. Up until the late 1800s abortion was a taken-for-granted facet of life. While sex out of wedlock was frowned upon and, in some places, forbidden, it was commonly understood that midwives and women within their own communities had means of inducing abortions to end unwanted pregnancies. As we considered in Chapter 2, before 'quickening' – that is, the point when a woman can feel the fetus move within her body, some time after eighteen weeks – these practices were permitted without being regarded as particularly controversial.

The introduction of criminal sanctions for abortion in America and Britain, beginning around the late 1800s, was expressly driven by white supremacist and misogynistic ideology. In both Western Europe and the Americas, amid the first murmurings of eugenics to improve the quality of the population, fears of a declining birth rate among wealthier white people prompted abortions to be viewed in a different light. At the same time the mostly male medical profession sought to wrest authority from midwives and other community caregivers, who continued to be the primary supports for women through birth, pregnancy and its prevention. Criminalising abortion, as intended, had the dual effects of bringing reproductive care under the control of male physicians and reinforcing the role of upper-class women as wives and mothers. The one aspect of early bans on abortion procedures in the United States and England that might be perceived as benevolent was that abortion could be dangerous, and botched abortions could cause serious harm. It was never the case, however, that *all* abortion procedures caused harm, and criminalisation has the effect not of making abortion safer, but of driving it underground. Making abortion the business of criminal law was always about power.

In that overheated conference room in 2018 the bioethicist who explored the idea that artificial wombs would mean that abortion could no longer be permitted was simply doing what bioethicists do. He was conducting a philosophical thought-experiment, imagining what might happen to abortion rights if ectogenesis existed, and not necessarily saying that he thought abortion should be banned. But in a material, non-hypothetical way what he was exploring – the idea that women might not be allowed to have an abortion if the fetus could live and continue to grow without depending on their bodies, and that fathers could be given equal rights to the fetus – are arguments that have already been used successfully for many years to undermine access to abortion, to criminalise women and to create barriers to care that have caused real injury. And this is without artificial wombs. The thought-experiment assumes there is a 'we' that collectively agrees that abortion should not exist. The assumption is that everyone would agree that abortion is a moral quandary. Even those who identify as pro-choice, it is presumed, would wish to do away with abortion if it were possible to find an alternative that would still protect bodily autonomy.

Contrary to the flippant commentaries of pundits who speculate on a future without abortion, many of us see abortion as essential healthcare, not as a political issue or a moral ill. If you have been involved in campaigning for abortion that is free, safe, legal and accessible, or even if you are someone who simply believes that people should have a right to choose, then you are unlikely to see this technology as the conclusion to the fight over abortion. Instead the end of that fight would be a reality in which we all have autonomy over our own bodies. It would be one where we are free to access all forms of reproductive care without risk of criminalisation or injury, and where a person making reproductive decisions on their own terms is not a topic of political debate. The end of the abortion 'debate', in other words, would be arriving at a place where abortion was no longer stigmatised. It would be abortion no longer being considered the business of criminal law. It would be reproductive care being accepted as healthcare, to which our access was so securely protected that we could simply take it for granted. It would be abortion no longer being treated as a topic worthy of a scathing op-ed, protest or platform.

From a feminist perspective that centralises the health and self-determination of pregnant people, the idea of artificial wombs as a welcome alternative to abortion poses some very apparent problems. First and foremost, as the legal scholar Emily Jackson has aptly observed, 'where abortion is illegal or unavailable, women do not continue their unwanted pregnancies'.[3] In places where abortion is illegal or hard to access, pregnant people still find ways to terminate, but they do so while being confronted with serious emotional, financial, physical and legal barriers. Those who argue that ectogenesis could 'end' abortion are assuming that anyone who set out to end a pregnancy would find the option of having the fetus transferred into an artificial womb equally acceptable. So long as pregnancy can begin in a person's body, though, there will be people who wish to terminate.

Imagine this: you have just discovered that you are pregnant. You think you might want a baby some day, but right now you know this isn't for you. You are told that if you want to end the pregnancy, you can, but the fetus will be removed to continue to grow in an artificial womb. This is not what you had in mind: you do not want a baby that began in you to continue growing alone, somewhere else. What if they contact you some day, wanting to know why you let a machine gestate them? What if something terrible happens while they are gestating in the artificial womb and you feel responsible? You aren't in a happy relationship with the fetus's other genetic parent. What if, while it is gestating, they make a claim to parenthood? How would you navigate the mutual friends, and peers, who might be involved in the fetus's care while it is gestating? What if you run into the nurse who assisted in your fetal transfer at the grocery shop: is she still working with the fetus that began with you and is now gestating in an artificial womb?

Behind the idea that artificial wombs will eliminate both the justification and the need for abortion is the assumption that the only legitimate motivation for seeking this procedure is a desire to physically end a pregnancy. Regardless of what a person ultimately chooses, what is being weighed when they make this decision is much more complicated than 'Do I want to be pregnant or not?' There is no one universal reason why people seek an abortion. More importantly, to understand abortion as vital healthcare, not as a moral problem, is to refuse the idea that a person should have to provide a justification for choosing abortion to begin with. What makes gestation that occurs in a pregnant person's body fundamentally different from ectogenesis is that whether wanted or not, pregnancy involves a person's physical and emotional experiences. A legal framework that took an affirming approach to abortion as healthcare would be flexible and adaptable, shaped by an awareness that different circumstances produce different needs. Sarah Langford, a feminist legal scholar has pointed out that the idea of artificial wombs as a 'solution' to abortion is incredibly dehumanising for the way it frames pregnant people as 'foetal incubators rather than people'. It is, she writes, 'assumed that the foetus can simply be transferred from one incubator (a woman) to another (a fake womb)'.[4] To understand the pregnant person's body and the artificial womb as exact equivalents, you must strip away the pregnant person's desires, needs and interests.

This is where the idea that anti-abortion and pro-choice advocates alike would be unified in welcoming artificial wombs as an alternative to terminating the fetus falls flat. At its heart, this argument is based on a very limited understanding of what 'choice' and 'autonomy' mean in practice. Societies that protect abortion rights do so on the grounds that being able to get an abortion is a fundamental part of being able to exercise autonomy. Those who propose artificial wombs as a welcome alternative to abortion argue that by allowing the pregnant person to remove the fetus from their body, the technology would protect their right to autonomy (by ending the unwanted pregnancy) while also preserving the fetus's life. But these concepts – 'autonomy' and 'choice' – can be understood in different ways. In a dictionary sense, 'autonomy' means self-governance, independence or freedom. Choice is the exercise of autonomy, the ability to choose the course of action that works best for you. In arguments that understand artificial wombs as a tool to end the justification and the need for abortion, autonomy is understood only in relation to the body: a person can enact a choice to be pregnant or not to be pregnant. Here, bodily autonomy is understood as the ability to end that pregnancy or continue it. These arguments imagine that once the fetus is no longer dependent on the pregnant person's body, that person's autonomy is no longer implicated. But even if we understand autonomy as only about being able to choose what happens to one's body, there is still a significant chance that requiring someone to have a fetus removed into ectogenesis, rather than terminating it, would be a substantial bodily violation.

From the 1970s to 1990s, many abortions were performed through dilation and curettage, a minor surgical procedure, or through vacuum aspiration. As of the 2010s, most abortions are medication abortions and occur very early in pregnancy. As noted earlier in the chapter, this means that many terminations involve taking two pills over the course of twenty-four hours, followed by a subsequent period of recovery that can be managed at home, with few exceptions. The idea that a procedure to extract a fetus into an artificial womb would essentially be equivalent to having a pregnancy terminated is deeply anachronistic. The artificial wombs currently in development are for fetuses at twenty-two to twenty-four weeks' gestation. And we can assume that even if that threshold were lowered over time to twenty, seventeen or fifteen weeks, a procedure to remove a fetus intact from a person's body and transfer it into an artificial womb would be significantly more invasive than taking two pills. It would require a pregnant person, presumably, to continue to gestate until an artificial womb could safely sustain the fetus. There is a clear contrast then, in terms of protecting bodily autonomy, between ingesting abortifacients in the early stages of pregnancy and being made to carry a pregnancy for weeks until a transfer became possible, at which point the fetus would likely need to be delivered or surgically removed.

For the sake of argument, let's assume that it would be possible to reach a point in the future where these two procedures were in fact equivalent. Abortion involving termination would require taking two pills, and a 'fetal transfer' would be no more invasive. Even then, the 'artificial womb as abortion alternative' argument still offers a very reductive view of what 'autonomy' and 'choice' mean in practice. To truly protect a pregnant person's self-determination would be to grant them the rights and the resources to choose what worked best for them, among a variety of reproductive options. To protect their access not only to an abortion, but to the kind of abortion procedure that was acceptable to them. As Emily Jackson writes, the autonomy sought in the abortion decision is not simply about 'a right to be free from unwanted intrusion, but ... the idea that individuals should be able to pursue their own goals according to their own values, beliefs, and desires'.[5] If ectogenesis were presented as the only option when someone sought an abortion, this would deny them the resources with which to decide according to their own 'values, beliefs, and desires'. None of this is to say that there is *no* possible scenario in which someone might prefer to transfer a fetus to an artificial womb – and have another person, or persons, take over its care. Providing an artificial womb as an option, however, would not be the same as mandating that you must use the technology if you wished to end your pregnancy.

In our inequitable world, even leaving open the 'option' of fetal transfer, instead of abortion involving termination, could result in significant harm. Given what we have explored in the last few chapters regarding both the challenge of creating an effective artificial womb for neonatal care and the likely cost and limited availability of any ectogenic technology, the idea that artificial wombs could entirely replace the need for termination requires a significant speculative leap. It is extremely unlikely that artificial wombs intended to treat wanted, prematurely born babies would be so widely available that every person who sought an abortion could use one. If transferring a fetus to an artificial womb was eventually presented as simply one alternative alongside termination, then the most likely case scenario is a hierarchy of access.

Even in jurisdictions like Canada where abortion is legal throughout pregnancy, access to abortion care is not equitable. Several provinces have scant services available, and people in rural and remote areas must travel long distances to reach providers. Abortion access, then, is already determined not only by legal restrictions, but also by geographical and financial ones. If artificial wombs were provided as an option in addition to termination, we would probably see a replication of already existing patterns in uneven access to care. In jurisdictions with anti-abortion legislatures, artificial wombs might be made available without also providing access to terminations. People with resources could find the means to end their pregnancies on their own terms, while others might find themselves forced to 'choose' between using an artificial womb, travelling in search of a termination or risking harm or criminalisation for self-aborting. People who are over-policed and under-cared for by the medical system – including Black and Indigenous women, women of colour, women from lower socio-economic backgrounds, young people, queer people and trans and non-binary pregnant people – are already most likely to face obstacles to abortion access. Artificial wombs presented as an 'alternative' to abortion would constitute yet another obstacle. They would exacerbate already existing conditions under which those with the most resources might have increased choice about what kind of abortion they wished to receive, while those with the least resources might find themselves coerced into either continuing their pregnancy or using an artificial womb.

And then of course there is the question of what exactly would happen to these ectogenetic children, whether they were removed willingly or coercively from people who did not want them to be born. Who would be responsible for these infants? Would they be placed up for adoption, and at what point in their gestation would a prospective parent need to be found? Where would the artificial wombs in which they gestated be kept, and who would be accountable if something went wrong? There is a palpable disdain for pregnant people inherent in the assumption that gestating a baby is as straightforward as simply finding a vessel for it. Commentators who suggest that if a pregnant person does not wish to continue a pregnancy, their fetus could simply be transferred to an artificial womb often remain suspiciously silent about the rather significant question of who, or what, would take responsibility for them. The fact is that whether a fetus grows inside a person's body or inside an artificial womb, it will never be an independent entity. The thought that artificial wombs would enable fetuses to be treated as autonomous beings, as persons with full rights, is unfortunately quite symbolic of a culture that has long been dismissive of the often-gendered work that goes into caring for babies.

Why is it so easy for commentators on the artificial womb to assume that a fetus could be an independent human if only the pregnant person's body was not in the way? A good example of this logic comes out in the work of two bioethicists, Peter Singer and Deane Wells, who are often credited as the first scholars to make the argument in the 1980s that artificial wombs will end abortion. They wrote that by mandating extraction to an artificial womb, 'abortions could be done using techniques that would not harm the fetuses, and the fetuses, or newborn babies as they would then be, could be adopted'.[6] For Singer and Wells, the fetus would be a 'newborn baby' – a legal person – if it were removed from the mother's womb. But even if it were inside an artificial womb, a fetus would not yet be a baby.

A fetus is still developing the capacity to survive outside the uterus. Its existence relies entirely on being gestated. When it grows inside a person, that person is providing everything it needs in order to continue to live. And when it is inside an artificial womb, that care is provided not only by the technology, but also by the people (whether nurses, doctors or technicians) who monitor it. We can imagine that at some point in the future, artificial wombs might be capable of entirely replicating gestation, but it remains likely that some form of human oversight will always be required. Who would look after these ectogenetic fetuses while they grew?

The World Health Organization estimates that forty to fifty million abortions occur worldwide per year. That is 125,000 abortions a day, that we know of. The real number, given the continued necessity of extra-legal abortion, is likely to be much higher. We cannot simply assume that the answer to the vast problem of managing the care of every fetus unwanted by a pregnant person is that someone would want to adopt them. For every situation where the other progenitor, or a family member, might come forward, there could be a situation where no one could be found to care for the infant, once it reached term. The infrastructure that would be required to gestate every unwanted fetus through ectogenesis would be costly, both in terms of machinery and in terms of human labour. Would they be wards of the state, or would a caretaker have to be found for each fetus before a transfer could occur? That each of these emotionally charged questions is frequently left out of the blithe claims that artificial wombs will replace abortion is indicative of how careless many of these authors are about what it takes to gestate a baby and to care for an infant, once born.

And how careless, too, about what a person must weigh up when they choose whether to proceed with a pregnancy. Assuming that these are considerations that would simply fall into place speaks to how little our society acknowledges that pregnancy, and then parenthood, is hard work – and work often done by one person alone, without resources. The idea that we should simply assume that fetuses in artificial wombs would be provided for, when so many pregnant people and parents are left to struggle, is a cruel contradiction. In a world where the children of migrants, of Black and Indigenous parents, of working-class parents are not only left without support, but often targeted with violence, it is unconscionable to simply assume that creating an infrastructure to grow otherwise-aborted fetuses through ectogenesis would be a straightforward social good.

In Britain in 2020 the Tory government was perfectly happy to deny funding to support free lunches for school children who would go hungry without. In America that same year at least 545 children remained separated from parents who were deported, and subsequently lost, when the state failed to record their information. In the very same month that the Supreme Court ruled to overturn *Roe*, sanctioning forced pregnancy, an infant formula shortage left people across the United States struggling to feed their babies. Given the incredible ways that the state continues to harm children and parents in minoritised communities within some of the world's wealthiest nations, preserving the lives of fetuses that would otherwise be aborted by transferring them into artificial wombs is not about respect for life, or 'saving' babies. If this was the case – if this was what was really valued by contemporary anti-abortion campaigners who might welcome the artificial womb 'solution' – then these groups would be out in droves fighting to protect the already existing babies whose parents are denied the means to ensure their survival, or who are the victims of state brutality. Instead this is, and has always been, about controlling gendered bodies.

It should be enough to consider all the ways that artificial wombs as a forced alternative to abortion would undermine the reproductive self-determination of pregnant people. However, this idea should also give pause to those who seem solely interested in the potential life of the fetus. While a fetus is not a person, it can become one. How would you explain to a born ectogenetic baby that they did not have parents, because their parents had been coerced into having them transferred into an artificial womb after seeking an abortion?

In 2018 the Turnaway Study, conducted by researchers at the University of California San Francisco, released the results of a longitudinal project started in 2010 on American women's experiences of being denied an abortion. The researchers estimated that around 4,000 women in America a year are turned away from abortion services. One striking finding was that 90 per cent of these women chose against adoption. This reflects what we should already know: putting a baby up for adoption is not equivalent to abortion. It might also be illustrative of how people seeking an abortion would feel about the alternative of an artificial womb, which, after all, would be more similar to adoption than to termination. But for those people denied abortion who subsequently have a baby who is adopted, does that adoptee then have a right to know their origin? Across the United States adult adoptees have organised for rights to access information about their histories that many people raised with their families of genetic origin take for granted. This includes knowing who your birth parents are, and whether you have surviving relatives who might wish to know you. It also includes access to your birth certificate and adoption records, and to aspects of your family's medical and social history that could potentially impact upon your life; plus knowledge of the culture that shaped your biological relatives, of the circumstances of your adoption, of generational trauma, of heritable conditions that might affect you or your children. Some people may not wish to encounter this information. Given the often complex and sometimes coercive circumstances that can surround adoption, however, it is unfair to assume that will be true for all adoptees. Would children born from artificial wombs be told that their genetic parent was given the choice between pregnancy and fetal transfer? Imagine how it would feel to search for information about your origins, only to learn that you were ectogenetically gestated after your genetic parent was denied a termination.

It is tempting to simply dismiss legal scholars, newspaper commentators and politicians who promote the idea of artificial wombs as an alternative to abortion. If, for all the reasons we have explored, this would never be acceptable from a perspective that affirms abortion as healthcare, then why even entertain their arguments? Regardless of how abhorrent these ideas are, these authors are unfortunately right in the sense that artificial wombs could pose a challenge to the *legal* protection of abortion in places where abortion remains governed by criminal law or is articulated in regulation as anything other than a vital human right. This does not mean that, after the technology is introduced, we should simply accept a limit to reproductive rights. Instead it means that, in many nations, abortion is insufficiently protected.

Abortion is not treated the same way in law across all contexts, and a particular combination of factors is likely to determine whether artificial wombs could threaten reproductive rights in a given jurisdiction. Those factors are whether abortion is regulated by criminal law, whether it is limited in balance with a fetus's presumed capacity for independent life, and what the prevailing political, cultural and legal attitudes are towards abortion. In places where abortion is prohibited entirely, such as the Dominican Republic, Iraq and Egypt, the artificial womb may have little impact on abortion rights, because no such right is yet protected in law. But in places where abortion is permitted under certain circumstances yet still regulated through criminal law, artificial wombs could pose a threat to its legal protection.

Artificial wombs are anticipated to be ready for human trials within a matter of a few years. Writing this in spring 2022, I speculated that even if fetuses growing in artificial wombs were not found to be viable, the balance of interests in US law between a pregnant person's privacy and fetal life could still allow states to use the technology to restrict abortion. In *Planned Parenthood v. Casey*, the Supreme Court re-emphasised that states could show interest in the fetus's life throughout pregnancy, even in the early stages, by initiating state-level restrictions to abortion. To articulate this balance, the court established the 'undue burden' standard, finding that 'undue burden exists, and therefore a provision of law is invalid, if its purpose or effect is to place substantial obstacles in the path of a woman seeking an abortion before the fetus attains viability'. In other words, states could restrict abortion prior to viability, so long as those restrictions did not place an 'undue burden' on the pregnant person. To ban abortion entirely before a fetus was viable would be to place a substantial obstacle in the path of a person seeking the procedure. But even the possibility of a fetus being able to survive in an artificial womb could mean that an anti-abortion state legislature could ban abortion causing the death of the fetus. If an artificial womb were available in principle, even if not always in practice, this kind of ban might not have been legally considered a substantial obstacle, even early in pregnancy, because in theory a person seeking abortion would have the option of having the fetus extracted to an artificial womb. This would be in keeping with the measures already taken in many states to make it near impossible for pregnant people to access local abortion services in practice.

These issues became all the more pressing after the appointment of Amy Coney Barrett in 2021, at which point the US Supreme Court became a 6:3 majority of sitting justices who are anti-abortion. In April 2022, the Supreme Court was poised to determine the direction that the legal right to abortion would take in America. *Dobbs v. Jackson Women's Health Organization* dealt with several measures taken in Mississippi to make abortion near impossible to access, including a ban on all abortion after fifteen weeks. When the case came before the Supreme Court, the conservative majority swiftly overturned the abortion right as carved out in *Roe v. Wade* and affirmed in *Casey*. As the dissenting justices acknowledged, the court had never truly recognised a fundamental, absolute right to abortion in the United States. It had upheld, instead, a balance of rights between pregnant person and fetus. With the Dobbs ruling, Justices Sotomayer, Kagan and Breyer wrote in their opinion, 'the Court discards that balance. It says that from the very moment of fertilization, a woman has no rights to speak of.'

A threat to abortion prompted by artificial-womb technology

in future is neither an acceptable possibility nor an inevitable one. It is important, however, to understand the way that reproductive rights in places like the United States have been eroded continually over time, and to be realistic about the fact that, in some places, artificial wombs are highly likely to be co-opted to further this cause. The Dobbs ruling is a glaring example of the dark path that technologies like the artificial womb could take if we do not pay attention. Understanding this before the technology arrives is what allows us to strategise to protect reproductive rights. In nations where abortion is permitted within carefully specified parameters, where pregnant people and medical practitioners can be criminalised for abortion outside these guidelines and where there continues to be strong anti-abortion sentiment, the technology poses a legal challenge.

This does not mean, as some commentators would have you believe, that artificial wombs pose a universal test to abortion rights. Even within the US there are states with broadly pro-choice legislatures. In California, Oregon and Connecticut abortion continues to be permitted throughout pregnancy, and in New York reproductive justice advocates and activists successfully pushed for the passage of the Reproductive Health Act in January 2019, which strikes down criminal consequences for abortion and allows for terminations to be self-managed. By the time this book goes to print, activists, politicians and healthcare providers across the United States will have engaged diverse strategies to reject and attack the injustice of the *Dobbs* ruling, and to protect access to care. There is significant variation as to how abortion is regulated in law across jurisdictions, and broader socio-legal and political attitudes are also important in determining whether artificial wombs are likely to undermine reproductive rights.

For example, in England, Scotland and Wales under the

Abortion Act 1967, 'procuring a miscarriage' remains a criminal offence if it is done outside the specific guidelines outlined in the legislation. While people might not think of abortion in Britain as something that remains associated with criminal law, the Abortion Act 1967 did not legalise abortion, but instead established exceptions to its ongoing criminalisation under several archaic pieces of legislation. Abortion has been criminalised in English common law for many decades. Section 58 of the Offences against the Person Act, which was introduced in England, Wales and Northern Ireland in 1861, established the statutory crime of a woman 'unlawfully administer[ing] to herself any poison or other noxious thing' with intent to procure a miscarriage, or of aiding another person in obtaining a miscarriage. In 1929 Section 1 of the Infant Life (Preservation) Act was introduced alongside the Offences against the Person Act to establish an offence for aborting a fetus 'capable of being born alive', unless done in good faith to preserve the life of the pregnant person. The Abortion Act 1967 establishes that a pregnant person will not be guilty of an offence if their pregnancy is terminated by a registered medical practitioner, and if two practitioners agree in good faith that the pregnancy hasn't exceeded twenty-four weeks. They must also agree 'that continuance of the pregnancy would involve risk, greater than if the pregnancy was terminated' of injury to the pregnant person's mental or physical health or that of her children; or that the termination is necessary to protect the pregnant person's health or life; or, finally, that there is a significant risk of the baby being born dead or with substantive anomalies.[7]

When the legislation was passed in the 1960s the upper gestational limit for abortion, after which permissions became more strictly regulated, was twenty-eight weeks. After this point, abortion was still permitted, but only on the last two grounds. In

1990, after a protracted debate over whether and to what extent improvements in neonatal care meant that the threshold for fetal viability had shifted, Parliament found in favour of lowering the gestational limit to twenty-four weeks.

It is possible that in Britain the artificial womb's impact on medical viability would lead to the gestational limit of twenty-four weeks being revisited and lowered once again, so that abortion would be more tightly restricted at earlier stages in pregnancy. Given that two medical practitioners must affirm that the pregnant person meets the requirements outlined in the Act, it is also possible that some practitioners might not agree in good faith that transferring the pregnancy to an artificial womb would be more damaging to the pregnant person than terminating the fetus. And even if medical practitioners remained largely in support of protecting abortion access, it is also possible that they might face a risk of criminal charges, if an argument could be made that transferring the fetus to an artificial womb would pose an equal or lesser risk than continuing a pregnancy.

This is where a different social and political ethos around abortion is likely to have an impact. In Britain a broad majority of medical professionals and politicians are vocally in support of decriminalising abortion and granting greater autonomy to pregnant people in the abortion decision. The long-standing and clearly articulated regulatory framework for abortion in England, Scotland and Wales means that while anti-abortion sentiment remains in some quarters, abortion is broadly treated as a settled political issue. Unless there is a significant change in the UK social and political context (which is of course possible), it is unlikely that we will see large-scale support for campaigning for the use of artificial wombs in lieu of abortion. While both Britain and the United States continue to have abortion jurisprudence that would allow for artificial wombs to pose a *legal* threat to abortion rights, the different climate in the UK makes the success of such a challenge less likely. Having said this, leaving open the possibility that artificial wombs could put pregnant people and medical practitioners at risk of criminalisation for abortion – even if this outcome is improbable – is clearly unacceptable. In nations that continue to treat abortion as an issue for criminal law, the possibility that artificial wombs could undermine reproductive rights is always on the table.

In 2020, during the Covid-19 pandemic, healthcare providers and charities, including the British Pregnancy Advisory Service, successfully lobbied the government to allow people to access abortion pills via telemedicine. The Abortion Act 1967 requires all abortions to occur in an approved healthcare facility. To move to allowing abortion by telemedicine, the government temporarily redesignated the home as an approved healthcare setting. While advocates were hopeful that this change would stick, in 2022 the government announced that it would revert to the pre-pandemic protocol, wherein people must have a consultation with a clinician and take the first pill in the two-pill cycle at a hospital or clinic. Wales has chosen to retain the pandemic changes. Reluctance to do so in England, as of 2022, speaks to the ongoing presence of a political and social contingent that continues to use the management of abortion access as a means of exerting control over gendered reproductive bodies. And people who find themselves facing unwanted pregnancies continue to be at risk under the current legal framework. As of summer 2022, two women who terminated their pregnancies using pills in England are facing criminal charges for abortion, and could face up to a life sentence. According to the British Pregnancy Advisory Service, at least seventeen women have been investigated for abortion over the last eight years. The treatment of abortion as something for which people should be punished if they do not follow the specific stipulations of the state is patronising and spiteful. So long as abortion is considered distinct from other forms of healthcare, access will remain vulnerable.

Over the years, legal scholars and bioethicists have proposed various means of rearticulating limited protections to abortion in law to maintain the status quo after artificial wombs. These proposals include redefining viability to refer to the fetus reaching the capacity to survive without any intervention; applying a property right or right to genetic parenthood; and extending a defence of abortion based on bodily autonomy, so that it could still apply after ectogenesis. Any of these might be used as legal strategies to protect the status quo in places where abortion remains criminalised. While each of these strategies might be effectively mobilised to offer some legal protection to reproductive rights where artificial wombs might threaten them, each also risks replicating the limitations that leave these rights open to attack in the first place.

The problem is not artificial wombs. Instead the problem is that in some nations, abortion continues to be constructed as something that requires a legal defence in the first place. The problem is the legacy of paternalism and fear of women having power over their own bodies on which these laws are built. We do not need limited legal solutions that continue to carve out a narrow space for reproductive freedom, or that accept the treatment of abortion as the business of criminal law. We do not need a law to govern our reproductive lives. Instead we need to decriminalise abortion, treat it as vital healthcare, and ensure that access that is safe, local and culturally competent is protected around the world. Where abortion is affirmed in this way, technologies such as the artificial womb, designed to support the health of wanted prematurely born babies, need not pose any challenge to reproductive rights.

Canada is a useful example here. While there continues to be vocal organising against abortion, and access is unacceptably limited in many parts of the nation, terminations are legal throughout pregnancy. At the federal level, potential for fetal life has never been articulated as a competing interest with the pregnant person's rights. The case that established the right to abortion in Canada, *R. v. Morgentaler*, came before the Supreme Court in 1988. Dr Morgentaler, a physician based in Quebec, and his colleagues, had been providing terminations in defiance of section 251 of the Canadian Criminal Code. Section 251 treated performing an abortion as an indictable offence, except in cases where a hospital committee agreed that continuing the pregnancy might result in a woman's death or damage her health.

Morgentaler had already found himself in court once. Like *Roe*'s legal team, which had sought to bring the criminalisation of abortion before the courts, Morgentaler and his collaborators hoped that if they could make a case to the Supreme Court, the prohibition would be found to be unjust. The court ultimately agreed that the requirement for committee approval violated women's rights under section 7 of the Canadian Charter of Rights and Freedoms, which holds that 'everyone has the right to life, liberty, and security of the person, and the right not to be deprived thereof except in accordance with the principles of fundamental justice'. Three majority opinions were read, with three different justices offering their interpretation of precisely how the restrictions violated the Canadian Charter. Of the three majority opinions, Justice Bertha Wilson – the only woman on the court at the time – gave the opinion that is the most 'often cited and used to support legal rulings on reproductive choice in other areas of law'.[8] Justice Wilson's opinion, and its frequent citation in case law to follow, articulates a broad protection for abortion in Canada. She wrote, 'the right to reproduce or not reproduce [...] is properly perceived as an integral part of modern woman's struggle to assert her dignity and worth as a human being ... [Under section 251(4)] she is truly being treated as a means – a means to an end which she does not desire but over which she has no control.'

The decision in *Morgentaler* established protection for abortion both as a right that was necessary to protect security of the person (that is, 'the physical and psychological integrity of the individual') and to protect liberty, which refers to the requirement that the state 'will respect choices made by individuals and, to the greatest extent possible, will avoid subordinating these choices to any one conception of the good life'. In Canadian law, the fact that abortion is legal throughout pregnancy means that artificial wombs would pose no immediate challenge to the protection of abortion. And the broader articulation of the abortion right as a matter of liberty also means that even if anti-abortion organisers attempted to bring forward a case against abortion with the introduction of the technology, they would likely struggle in court. A case could be made that requiring a pregnant person to have a fetus transferred to an artificial womb, rather than being terminated, would not be a significant infringement on the security of the person. But to require someone to use an artificial womb rather than terminating their pregnancy could indeed be said to be undermining a right to liberty by subordinating their choices to 'one conception of the good life' and challenging the pregnant person's 'dignity and worth as a human being'.

Canada is far from perfect in its protection of abortion. In many parts of the country, all forms of reproductive care remain difficult to access. In some provinces, like Alberta and Manitoba, abortion services are available in cities, but not in remote or rural areas. In others, like Nova Scotia, Prince Edward Island

and New Brunswick, services are not available after the first twelve to sixteen weeks of pregnancy. In the rare cases (less than 2.5 per cent of all abortions) where people need access to terminations after twenty weeks, they must travel to one of three cities where such services are available, and in some instances travel to the United States. This lack of accessibility, and the requirement that people sometimes travel and pay out of pocket for abortion, is not good enough. As we have considered in this book, it is never the case that a legal right to something is adequate on its own: for rights to be meaningful, they need to be accompanied by resources and support that enable people to actually exercise them in practice.

What the Canadian example shows, however, is that where abortion is not associated with sanctions in criminal law, artificial wombs do not necessarily pose a threat to its *legal* protection. Abortion jurisprudence in this context protects pregnant people from criminalisation. This approach to abortion in law – as healthcare, rather than as stigmatised, criminalised procedure – creates the necessary minimum conditions under which artificial wombs are unlikely to result in subsequent legal restrictions on abortion. We need something beyond the protection afforded to reproductive rights in the Canadian context. We need the affirmative protection of broad abortion access, and services that are localised, culturally competent and centred around protecting pregnant people's self-determination.

In July 2013 a thirty-three-year-old woman arrived at the emergency room in South Bend, Indiana, where she lived and worked at her family's restaurant. Not long before, she had messaged a friend who was a nurse to ask about pills to induce abortion. At the hospital, bleeding profusely, she told doctors that she had had a miscarriage. Scared, in pain and seeking help, she explained that she had not known what to do and had left the remains in a dumpster. Her honesty with the medical personnel she encountered that day would ultimately lead to police examining her phone records, collecting the fetus and incarcerating her under a 2009 Indiana state statute. The primary physician that she spoke to was avidly anti-abortion and it was his call to the police, and his trip accompanying them to the site where she had left the fetal remains, that initiated her criminalisation. She stood trial on two (conflicting) charges: feticide and child neglect. How could a person stand accused of both killing a fetus in the womb and of neglecting a born infant?

In a drawn-out court case, the defence insisted that the statute, which had been passed with reference to cases where pregnant women were subjected to violence that led to the death of the fetus, was never meant to criminalise pregnant women themselves. The defence also explained that the fetus, estimated to be twenty-three to twenty-four weeks, was stillborn, and consequently their client could not be found guilty of feticide. Prosecutors tried to argue that she had attempted to induce abortion, but the fetus had been born alive. To defend their case, they used a 'lung float test' created in the seventeenth century as a means of identifying infanticide. The dubious logic of this test, which was discredited in the early 1900s, holds that lungs that have respirated will be aerated, allowing them to float in water, while lungs that have never been exposed to air will sink. The prosecutors alleged that because the fetus's lungs floated, it must have been born alive, breathed air and been left to die. Though no self-respecting scientist would credit the test's results, it was accepted as evidence that the fetus was past viability and had had a chance of survival.

Under the Indiana statute, the charge of feticide could still apply for attempting to terminate a fetus, even if this termination was unsuccessful. Although in Indiana abortion pills like the ones this woman took could be prescribed by a doctor, she was found guilty of feticide for administering them to herself. And because the prosecution alleged that her attempt at feticide failed, she was also charged with child neglect. She was sentenced to twenty years in federal prison. A year into her sentence, her legal team appealed her case, with partial success. The appeals court vacated the charge of feticide and reduced the length of her sentence on the charge of child neglect.

Her conviction happened in a state where, under *Roe v. Wade*, reproductive rights were protected by law. It illustrates, precisely, that it is not enough simply to have a limited legal right to 'choose' abortion without government interference. We need to have free, open access to both abortion care and all other forms of reproductive care, without stigma. And we need to be certain that pregnant people can access this care without being criminalised.

The limitations of *Roe* did not impact upon all pregnant people in the same way. A framework that protected a narrow legal right to choose abortion, but didn't require states to provide access in practice, acted primarily to protect white women with resources. With the fall of *Roe*, it will continue to be disproportionately Black and Brown pregnant people, immigrants, those with low incomes, and young people who are most likely to be criminalised and denied crucial care. An artificial womb, introduced into this context, would be no different, and any policy to require people to use an artificial womb rather than terminate would hit already marginalised groups the hardest.

The focus of both conservative commentators and liberal legal scholars on how artificial wombs might impact the status quo of abortion rights does a disservice to the many organisers, activists and progressive lawyers around the world who have long campaigned to reframe discussions of abortion, to understand it as healthcare and just one among many essential resources that are insufficiently protected and inequitably distributed.

Within individual nations, the kinds of measures that must be taken to secure abortion access before ectogenesis will vary, based on existing practice. However, two factors that are consistent, regardless of location, are disassociating abortion from criminal law and protecting access to all forms of reproductive care. The Guttmacher Institute estimates that approximately 20 million unsafe abortions occur each year, with the majority occurring in countries with the most restrictive abortion laws and low access to contraception. Restrictive abortion law does not mean fewer abortions; it means fewer safe abortions, and a greater threat to pregnant people's health and well-being. Rates of unintended pregnancies are also higher in countries that most severely restrict abortion, and lowest in countries that have broadly legalised abortion. In places that restrict abortion most, the number of unintended pregnancies ending in abortion has in fact gone up since the 1990s. Perhaps the strangest thing about all the discussion over whether artificial wombs will challenge legal protections to abortion is the assumption that the law should have any business delineating abortion rights at all. In the narrative of anti-abortion advocates, abortion has always been a criminal act, but that is a misrepresentation of history. The criminalisation of abortion is a relatively recent construct, and it can be undone.

We should decriminalise abortion everywhere, throughout pregnancy, without exceptions or gestational limits, because to do so centralises the lives, health, desires and needs of pregnant people; and not to do so places them in danger. But to speak directly to those who don't want abortion to occur at all: the statistics bear testimony to the fact that the best way to reduce unintended pregnancies, and thus reduce abortion, is to decriminalise it. Decriminalising abortion and making it available upon request without limitations, while certainly not the only measure that must be taken, is a step towards divorcing ectogenesis from the discourse on abortion, and towards creating space for exploring how the technology could benefit pregnant people without the spectre of how it could end their abortion rights. In turn, it would allow us to shift our emphasis to other ethical questions, such as when, and under what circumstances, people would have access to the use of ectogenesis.

Measures such as universal access to medication abortion through telemedicine would drastically improve care. An immediate response to the scarcity of abortion services in many parts of the world is ensuring that pregnant people will not be criminalised for seeking abortion pills via telemedicine and self-treatment. Campaigning for self-managed abortion, rather than repeating the refrain that is so common in writing on abortion and ectogenesis (that abortion must be regulated and justified), offers a new frame. It centralises the pregnant person, in not requiring them to seek approval or supervision from a physician to have a procedure that they themselves can manage. It allows them to determine the conditions under which their abortion occurs: alone or at home in the company of a loved one. It acknowledges the restrictions that shape and determine whether access is feasible: if a clinic is not available to someone, or if it is, but it will require them to undergo lengthy approval procedures – or if, for cultural or personal reasons, they do not trust the care they will receive – it enables them to secure abortion on their own terms.

Every time the researchers working on partial artificial wombs to save the lives of wanted preterm babies announce a breakthrough, a chorus of commentators excitedly speculate that the end of abortion rights is nigh. Writers and academics alike variously ponder whether this will mean that abortion is banned entirely, and float purported solutions to protect abortion rights against an imminent technological attack. Rarely is the question 'Why are we still having this conversation?' asked. By thinking of abortion as a 'debate', and of reproductive rights as reducible to a bare right to end a pregnancy, which might be taken away the moment innovation allows us to separate gestating from the body, these authors accept abortion as a contingent, moral issue.

We can instead reject the co-optation of a technology that could benefit the health of pregnant people – and of wanted, prematurely born babies – by an agenda that would hurt them. We should not take the fact that artificial wombs could undermine reproductive rights as evidence that abortion is a 'problem' to be solved by technology. Instead we should take it as evidence that protections to abortion are not strong enough to begin with. We should take it as a clear indication that it is long past time to decriminalise, destigmatise and improve access to safe, free and culturally sensitive abortion care.

6

The Tyranny of Biology

In her 1970 manifesto Shulamith Firestone decried pregnancy as an unconscionable social ritual. She had spoken to friends who had been through it, and their experiences affirmed that 'the temporary deformation of the body of the individual for the sake of the species' was patently unjust.[1] At the end of the day, as one woman candidly expressed it, giving birth was 'like shitting a pumpkin'. The symptoms of gestating included everything from inhibited movement, to nausea and death. Women were left with permanent physical injuries and trauma. And despite all this, they were still made to feel as if their worth was tied up in having babies. It should hardly be surprising that a woman might reject motherhood, yet childless women were still treated as unnatural.

They had been led to believe that no matter how horrific the complications of pregnancy might be, they should accept child-bearing as their inevitable lot in life. The joy of having a baby at the end of the process was meant to be enough to make women forget the cruelty of their experience and embrace the 'natu-ralness' of birth. But, Firestone argued, humanity had created innovations to improve on so many other things that had previously been accepted as 'natural'. We had successfully sent astronauts into space, yet we were still told there was nothing to be done about the fact that women bore sole responsibility for gestation. If men could give birth, all the world's resources would already have been directed towards creating an alternative to pregnancy. The problem wasn't a lack of technological capability, it was a lack of will. The fact that women had to bear the brunt of human reproduction was, to Firestone, a consequence of the 'tyranny of reproductive biology'. It might be an accepted phenomenon, but it was also a vast injustice that must be rectified. If we could automate gestation, she imagined, we might be able to change gendered relationships to reproduction. In so doing, perhaps we could make the work of childrearing a communal act, the responsibility of all of society rather than a burden carried by mothers alone.

Firestone would not have been especially impressed by contemporary research on partial artificial wombs. Back in the 1970s, she wrote about the antecedents of these kinds of technologies in her manifesto. The research in the 1950s and 1960s on artificial uteri, including the experiments by the Stanford professor who had run into trouble with student protestors, was, Firestone noted, far from the work towards ectogenesis that she had in mind. Expensive investments in keeping prematurely born babies alive, she argued, were fundamentally different from the possibility of feminist researchers creating a means of communally gestating children. To Firestone, it was typical of the patriarchal scientists of the time, 'few of whom [were] feminist, or even female', that any progress towards artificial wombs had to be justified on the grounds that it could save babies.[2] Liberating women was not seen as a legitimate aim for scientific enquiry in and of itself, and that was part of the problem. Firestone might have been dismayed, if unsurprised, to learn that in

2022 we are still a far cry from achieving a viable alternative to gestation. But the idea that ectogenesis could be a productive tool for 'liberating' women from shouldering an uneven burden in reproduction has fierce modern defenders.

In a boldly titled paper, 'The Moral Imperative for Ectogenesis', the philosopher Anna Smajdor proposes that 'the fact that women have to gestate and give birth in order to have children, whereas men do not' is a natural inequality.[3] Pregnancy, she goes on, requires women to be 'the sole risk takers in reproductive enterprises'. That is a significant burden when those risks include not only mild nausea and haemorrhoids, but also diabetes, pre-eclampsia, psychosis, depression and death. Like Firestone, Smajdor wholeheartedly rejects the idea that the only justification for creating an artificial womb is saving preterm babies.

She is in good company. A 2015 book by the Australian bioethicist Evie Kendal is devoted to the argument that the state should sponsor ectogenesis for the sake of equality between men and women and between fertile and infertile people. She too contends that the preservation of premature babies is neither the only, nor the most important, reason to fund research into artificial wombs. When it came to carrying a pregnancy, Firestone wrote, women were continually being led to believe that it was routine to be in discomfort and pain throughout the entire process, culminating in childbirth, which, she added, 'hurts. And it isn't good for you.' Today, when considered alongside preventing or ending gestation, carrying a pregnancy to term still remains the most dangerous option. As Evie Kendal writes:

pregnancy and childbirth are known to pose numerous health risks, with some 'normal' pregnancy-related symptoms including morning sickness, dizziness, headaches, bone and muscle aches, loss of visual acuity, bleeding gums, breathlessness, heartburn, varicose veins and hemorrhoids. Importantly, despite the significant discomfort many of these symptoms impose, as they are considered 'normal' they often fail to be acknowledged seriously and many are left untreated.[4]

For both Smajdor and Kendal, the fact that women continue to bear the brunt of pregnancy and birth, and that these experiences can be dangerous and painful, should be enough to justify resources being allocated towards the creation of ectogenesis.

Not everyone will agree that pregnancy and childbirth are so onerous that research into an alternative to gestation is warranted, or that ectogenesis is the ultimate solution. One thing that is undeniable, though, is that the sidelining of research into health concerns that primarily impact upon women has continued, from the time of Firestone's writing in the 1970s through to Smajdor's work in the 2000s. In 1969 Firestone wrote that 'the kinds of research done are only incidentally in the interests of women when at all'. Unfortunately, this remains all too relevant today. Pregnancy presents a host of compelling research questions. Contraception, miscarriage, chronic illnesses such as endometriosis and polycystic ovary syndrome, the multiple complications that can arise in the perinatal period, vaginal tearing, uterine prolapse, incontinence after birth and the symptoms of menopause are some of the many issues that primarily impact the lives of women. They are also grossly understudied and under-addressed. The purpose of biomedical research, after all, should be to understand and explore the dimensions of a problem and try to fix it.

Firestone contended in the 1970s that since scientific research was dominated by men, problems that impacted women were not even considered, let alone treated as legitimate concerns that warranted solutions. In one especially prescient example, she noted that studies on oral contraception for men in the 1960s were entirely dismissed, because it was easier to lay all the responsibility for managing reproduction on women. When researchers finally revisited the pursuit of oral contraceptives for men in earnest in the 2000s, outrage at side effects including fatigue, reduced libido, acne, mood swings and headaches brought the project to a screeching halt. As of the 2020s, ongoing research has been stalled in part by a lack of interest from pharmaceutical companies, which, after all, have a thriving business in the form of contraceptives marketed to women.

If you have been on oral contraceptives, you will find that list of side effects strikingly familiar. It wasn't until I had been on the pill for ten years that a nurse told me that, as someone who had suffered from migraines with auras, I was at an increased risk of stroke if I continued taking the combined oral contraceptive. The side effects of oral contraception that is readily available to women include acne, weight loss or gain, low libido, depression, anxiety and blood clots. When the AstraZeneca vaccine was limited in some nations due to a vanishingly small risk of clotting (approximately one in 100,000), a comparison used to encourage people to see that the shot was safe was the much higher incidence of blood clots posed by some oral contraceptives. Oral contraception is not fundamentally bad or dangerous. It is important to understand that the risks of clots from the pill are very low too. And while some women do experience side effects, many do not. The issue is not that contraceptives can have side effects. It is the hypocrisy of a society that decries any reproductive risk to men as totally unacceptable, but is perfectly happy to allow those risks to fall on women. It is the fact that little attention is paid to accurately researching, tracking and explaining adverse reactions to contraceptives that are primarily taken by women. It is the continued assumption that women will take sole responsibility for pregnancy and its prevention, and the lack of interest in the consequences of bearing that dis-proportionate burden.

The pattern that Firestone observed in the 1960s and 1970s is frustratingly persistent. At the time of her writing, *Roe v. Wade* had not yet been won. And while the pill was soon to become widely available, access was beset by legal barriers and fraught with ethical issues. John Rock and Gregory Pincus (the very same researcher whose work in the 1930s had helped inspire *Brave New World*) won the FDA's approval for the first oral contraceptive. Their first trials had been conducted on women in Puerto Rico and on patients at a Massachusetts psychiatric hospital, without informed consent. The earliest iteration of the pill contained substantively higher levels of hormones than the pills taken today, and consequently was much more likely to produce side effects. The physician overseeing the Puerto Rican trials advised in her initial report that 17 per cent of the women involved experienced unpleasant symptoms, including migraines, nausea and dizziness, and that these effects were too significant to justify the pill's use. Even after the FDA saw evidence of blood clots linked to the pill, it proceeded with the approval process, partly on the grounds that the dangers of oral contraception were still lower than the dangers women faced in pregnancy. Once the pill arrived in the United States, its use as a means of preventing pregnancy was still denied to many women.

In the 1960s the Supreme Court overturned a Connecticut statute that prohibited the use of contraceptives, ruling that the privacy rights of married couples extended to being able to use contraception as they chose. But it wasn't until the early 1970s that this ruling was extended to single people. When Firestone wrote her manifesto, then, a whole array of interrelated injustices was occurring around women's reproductive bodies. The pill, with all its potential as an empowering means of pregnancy prevention, was denied to some women, and was trialled on others with no concern for their health or their right to informed consent. These were two sides of the same project of controlling women's reproductive bodies. Even as the oral contraceptive was approved – despite its risks – on the grounds that it was safer than pregnancy, abortion access remained an uphill battle. While some women were able to secure safe abortions through feminist extra-legal providers, such as the Jane Collective in Chicago, others lost their lives or health seeking termination.

And then there were the consequences that followed from even much-wanted pregnancies. Firestone was horrified by how women were treated when they gave birth. In some hospitals it was routine for their lives to be secondary to those of their babies, for their wishes to be ignored, and for doctors to perform painful and often unnecessary procedures such as pre-emptive episiotomies. For years this remained a common practice in which physicians cut into the perineum (the tissue between the vagina and anus) before birth, in the mistaken belief that women would inevitably tear and that surgical incisions would heal better than a naturally occurring birth injury. The medicalisation of birth that had begun in the 1880s, as white male physicians sought to take reproductive matters out of the hands of women and midwives within communities, had calcified.

In response to these kinds of measures in the United States, 'natural' childbirth became a popular cause for the women's movement in the 1960s and 1970s. Encouraging women to embrace medication-free deliveries and give birth at home, rather than in clinics, was intended as a way of resisting the framing of pregnancy as an illness and the paternalistic mistreatment that women had been subjected to in hospitals. Firestone, however, viewed this movement with suspicion. To her, it was just another means of reinforcing the message that a woman's value was directly linked to her capacity for motherhood. Doctors would give women meds, but still expect them to tolerate negative outcomes as par for the course. Proponents of 'natural' birth would reduce women to their reproductive organs by insisting that childbirth was a source of feminine power and recommending against interventions that might make the whole experience less painful. Either way, there was a failure to see the incommensurate burden that pregnancy placed on women as an injustice that could, and should, be rectified.

Today it remains true that the over-medicalisation of birth, and its counterpart in the pressure to have a 'natural' birth, can be equally harmful to birthing people. Being pressed to give birth 'naturally' can make people feel as though if they need interventions such as a C-section, or request medication, they have failed. In 2022 a report into the deaths of nine mothers and the deaths or permanent injuries of more than 200 infants at the Shrewsbury and Telford Hospital NHS Trust cited an overemphasis on 'natural birth' as a key cause of these tragedies. Women were denied lifesaving procedures, including Caesareans, and were blamed for the loss of their babies. What happened to these families is an excruciating example of the consequences of forcing one concept of how birth should occur on pregnant and birthing people. In this instance an ideological attachment to the idea that labour should occur 'naturally', without intervention, resulted in deaths and injuries. By the same token, in other medical settings, particularly in North America, pregnant and birthing people report being pressured to undergo often unnecessary procedures, such as the induction of labour in the days after their due dates, with little concern for their wishes or for the effect of these procedures. Birthing people have been pressed, and even threatened, to accept interventions that are not medically required, sometimes resulting in significant physical damage and emotional trauma.

A hospital birth or a home birth can be equally empowering. A desire to labour with the assistance of an epidural or without any medications should be equally respected. People's preferences as to where, with which tools and with whom they wish to give birth should be heard. Respectful care during pregnancy and labour means supporting people to understand what is available to them, under what circumstances medical intervention may be necessary and lifesaving and what strategies they can use to feel empowered during birth. Taking these choices away – whether by insisting that someone's body is 'made' to labour and discouraging them from taking medication, or by demanding that they accept treatments they do not need or want – is unacceptable.

It would be unfair to say that nothing has changed since the publication of Firestone's book in 1970. The combined efforts of feminist scientists and researchers, as well as patient activists and reproductive health, rights and justice organisers, has led to substantive progress in research focused on reproductive health and pregnancy. It remains the case, however, that many of the complications of pregnancy, and many of the 'normal' aspects of it that can be painful, unpleasant and dangerous, are still woefully under-researched. Because of more stringent restrictions on studies involving pregnant people, too, we often have a gap in our knowledge of how medications and treatments affect them. The solution to this lack of data is that a precautionary principle is often applied that can create an additional set of concerns if you are pregnant.

For years, pregnant people have been advised not to take NSAIDs (such as aspirin or ibuprofen) during pregnancy, and to stick to acetaminophen (such as paracetamol) for pain relief. In

2021 researchers called for the 'precautionary principle' in the use of acetaminophen during pregnancy, citing a possible concern that it might impact fetal development. Pregnancy, of course, is a time when you are highly likely to experience pain: headaches, joint pain … you name it. And as it turns out, it also puts strain on your growing baby if you yourself are suffering. What exactly are we expected to do to manage? Grin and bear it? Notably, the precautionary principle in this instance meant advising that if people indeed had to take acetaminophen, they should take the smallest dose for the shortest possible period. Yet as I learned when a pharmacist told me on the heels of this study that 'it wasn't safe' for me to take Tylenol, the way this principle is often interpreted and applied is to recommend against consumption entirely.

As we considered in Chapter 3, when a food, beverage or exercise is understood – even anecdotally – to have potentially adverse effects in pregnancy, women are often simply told to abstain. I've learned while writing this book that it is a surprisingly common impulse for people to dismiss the idea that having to set aside certain activities or experiences during and after pregnancy is an issue. Indeed, bioethicists have cautioned that some measures would have to be in place to ensure women didn't seek to use artificial wombs for 'frivolous reasons.'[5] But who gets to say what is 'frivolous'? Cisgender men aren't typically expected to give up coffee, alcohol, cheese or fish for at least a ten-month period in order to have a child. Nor are they expected to ease off running, dancing or climbing. Nor do they have to experience nausea, gas, dizziness, mood swings or general fatigue, all to prepare to undergo the process of birth, plus potentially many more months of a baby relying on their breast milk. These months, too, often include being unable to enjoy activities that might have been a substantial source of joy before pregnancy.

As a habitual and happy drinker of two to three cups of coffee a day, I resented cutting back when I became pregnant, particularly given the lack of clear evidence that my consumption was in fact in any way problematic. I hated giving up wine too, and did not share the experience of many friends who told me they had suddenly found the thought of alcohol repulsive when their pregnancies began. Having moved through most of my life inhaling whatever chocolate, cheese, meat, uncooked vegetable or raw seafood was available to me, I struggled immensely with rethinking my food choices to ensure they were 'safe for baby'. And perhaps beyond anything else, the hardest part has been reaching the end of my pregnancy and realising that I simply do not have the capacity to do all the things that I have always enjoyed. The immense pain in my pelvis informs me that I will soon need to stop running. The mood swings and utter exhaustion at the end of hikes and walking my enthusiastic dog tell me that I need to do less activity and more lying down.

There is also the unanticipated way that birth hangs over me: what if something goes wrong? How can I really write down 'preferences' for a process I've never done before? How will I know if I want medication until I'm in labour? What's with all the 'advice' that turns out to be a roundabout way for someone to tell me a horror story about birth? How do I synthesise all the various techniques for breathing, the lists I've been given of things to keep in mind for the hospital, for post-partum, for feeding the baby, when I'm still trying to get through to maternity leave and grow a human? Those twin pressures of the medicalisation of pregnancy and the push to have a 'natural birth' also reverberate in the kinds of throwaway comments I have heard as my bump grows bigger. Am I going to take medication? Because really anyone can have a 'natural' labour if they simply learn to think of the pain as discomfort instead. I should simply let birth happen to me, because baby knows best. Will 'they' (meaning presumably the hospital? doctors in general?) 'let' me go past my due date or will I be induced? Do I know that I look huge? If the baby is that big, won't I need a C-section? Aren't I a little small for thirty-seven weeks and shouldn't I get that checked out?

The designation of 'frivolous' to any of the many possible reasons that a woman might find pregnancy difficult, frustrating, unpleasant or unfair is patronising. From the time of Firestone's writing in the 1970s down to the present, it has never been true that everyone experiences pregnancy in the same way. Many people describe gestating as a deeply happy time in their lives, and experience birth as empowering. For others, it is quite literally life-threatening. It is not possible to categorise pregnancy as fundamentally 'good' or 'bad'. I have resented equally the tendency to bypass all the challenges of being pregnant and the assumption that the whole thing must be a nightmare. It was a peculiar sensation, at the end of my second trimester, when I was generally energetic and happy and enjoying feeling my baby kick, to have people speak to me as though I was gravely ill and in need of rest. But we should be able to agree, without it requiring us to make sweeping generalisations, that preventing, ending and continuing pregnancy each present real risks – risks that can range from discomfort to death, and risks that continue to be shouldered primarily by women.

Aside from the physical and emotional challenges that can come with gestating, pregnancy and birth continue to be undervalued as real forms of labour. Women remain disproportionately responsible for childcare, an imbalance that often begins during the period of pregnancy, but continues long afterwards. As of 2022, the United States continues to be the only wealthy country that offers no guaranteed paid parental leave at the federal level. The 1993 Family and Medical Leave

Act (FMLA) offers a paltry twelve weeks of unpaid maternity leave, but 40 per cent of workers remain ineligible even for this sub-par coverage. While some states, like New York, DC and Oregon, have paid family-leave laws, in many parts of the US mothers and gestating parents find themselves needing to return to work two weeks post-partum, and fathers and non-gestational parents are unable to take any leave at all.

Countries including Estonia, Sweden, Lithuania and Hungary are world leaders in access to paid parental leave. However, many nations lag in providing paid leave for mothers, and in more than ninety-two nations no paid leave is available to fathers. Part of what Firestone saw in the 'tyranny of reproductive biology' was the way the expectation of motherhood curbed the possibilities available to women in determining the course of their lives. Firestone, an outspoken Marxist, would not have been particularly interested in policy changes focused on ensuring that women had an equal opportunity to be overworked by their employers. After all, it was a revolution that she was after. Liberal feminist proponents of ectogenesis as an alternative to gestation, however, frequently point to how pregnancy can impact upon women's economic status and career opportunities.

In 2015 Evie Kendal argued in her case for state-sponsored ectogenesis that 'the social burden of pregnancy is twofold: social expectation demands women become pregnant, and then pregnancy serves to materially damage women's social lives and limit their future opportunities'.[6] Today, even where paid parental leave is available, in places where it is limited to gestational mothers only, the expectation is that mothers, and not fathers or non-gestational parents, will take time away from work. When only women who give birth are granted leave, the message is that when women have babies they should stay home with them, while their partners continue with their careers. The issue is not simply limited to whether there is an available policy to protect paternity leave. South Korea and Japan offer some of the most generous paternity policies in the world, at more than one year's paid leave. But few men take this up. Persistent archaic norms around gender roles require that paternity leave also be incentivised through measures such as designated, or even mandated, leave for dads. The idea that a father in a heterosexual relationship would be the person to pick up the phone when daycare calls, or stay home with a sick child – let alone that he would care for a wailing infant – remains unthinkable.

The ongoing prevalence of the 'mommy track' or 'the motherhood penalty' means that after taking maternity leave, women may find themselves overlooked for opportunities and promotions. At the very same stage where men may find themselves working their way up or expanding their experience, women find themselves plateauing or being demoted. From the 1960s till today, the high costs of childcare and the fact that women on average are paid substantively less than men creates a pressure-cooker situation in which women in heterosexual relationships are often faced with the possibility that being a stay-at-home parent might be a more financially viable option than staying in work. It is also the case that in heterosexual relationships, even when both parents are in full-time work, women still perform the bulk of childcare labour.

When the Covid-19 pandemic led to school closures and virtual classes, working mothers shouldered the care. The uneven impact of childcare on women was reflected across all industries. In academia, for instance, where a strong track record of publications and successful funding bids is necessary to career progression, grant applications and journal submissions authored by women plummeted at the same time as they increased for men. The Center for American Progress reported that, across the nation, more women than men reduced their hours or left their jobs. After a long spring of lockdowns, one in four women in the US reported being unemployed due to unmet childcare needs.

These patterns are also racialised. In wealthy countries like the United States and Britain, it is often through outsourcing care and home labour to Black women and women of colour that white mothers with means are able to focus on career progression. More so than white women, Black, Latina and Indigenous women in the US were more likely to be in frontline employment during the pandemic and unable to leave their jobs, for financial reasons. These groups of women disproportionately needed to reduce or alter their work hours to manage childcare needs, creating additional financial strain. Kendal's argument that childbearing and rearing can undermine women's prospects holds, at least in some parts of the world. The question isn't whether pregnancy and care labour still disproportionately impact upon women. We know the answer to that question is a resounding yes. Instead the question is: would ectogenesis solve this problem?

The idea that artificial wombs could be used to redress gendered inequity in reproductive labour is compelling. This is perhaps because creating a technology that could automate gestation seems almost straightforward, next to fixing long-standing social messes and dismantling deeply engrained structural barriers. Unfortunately this is yet another example of looking to technological fixes to solve societal problems. An artificial womb is no substitute for a year, or even two, of statutory paid parental leave for parents of any gender. Nor is it a substitute for closing the gender wage gap or making childcare free for all. Nor would it provide support for single parents bearing the weight of pregnancy and care duties alone. And it certainly would not, on its own, resolve the lack of investment in research to tackle health issues that affect pregnant and birthing people. Commentators like Smajdor and Kendal are right to argue that the physical and emotional risks of pregnancy, and the devaluing of care labour, have a disproportionately negative impact on women. But when they suggest that the way to confront this is for society to invest in the development of artificial wombs they are also suggesting that the root of the problem is that women can gestate and men cannot. The problem doesn't lie with reproductive biology, however. Instead, sexism and medical paternalism are responsible for the risks posed by pregnancy not being taken seriously and properly addressed.

Turning to artificial wombs as the solution to these issues also has the unfortunate effect of placing the blame on women's bodies, rather than on the social and political problems that make it difficult to gestate and to be a mother. To be fair to commentators like Smajdor, those who make the case for ectogenesis often recognise that they are positing a technological solution to a social problem. It is possible to both acknowledge that substantive change is needed, and that progress is simply too slow. Women have waited for childcare fixes that work, for a genuine redistribution of the risks they disproportionately bear. Feminist commentators who write in favor of ectogenesis take the view that, at this point, social efforts are taking too long. Ectogenesis is a nuclear option, and one that is necessary.

In a sense, they are right. How can it be that fifty years on from Firestone's *Dialectic of Sex*, so many of the injustices she wrote about are still stubbornly in place? How can it be that many of us still face the same challenges in balancing careers and caring that confronted our own parents? It is true that it is taking too long to come. There is little reason, however, to believe that an artificial womb will come faster. As we have explored in this book, one constant throughout history has been that technologies can only be as progressive as the contexts into which they are introduced. Inequity in care labour has persisted since well before Firestone wrote in the 1970s. And it has persisted because it is built into the very fabric of our society. It is replicated through the aggressive reinforcement of binary ideas of gender in our children, and it is upheld by nearly every social institution on which we rely. If we had artificial wombs, we would be more likely to use them as yet another tool to bolster these norms than as a way to obliterate them.

These days, Firestone's name has cropped up in a handful of articles on ectogenesis, where she is often represented as a somewhat naïve technophile who was sure ectogenesis would empower women. In fact she proffered an important caveat to ectogenesis freeing anyone. She was interested in how the technology could be used after a social uprising, not in the absence of one. She wasn't calling for equity. She was calling for a radical rethinking of society, for a feminist insurgency that didn't just 'elimin[ate] male privilege' but also 'the sex distinction itself'. Utopia wasn't a reality where men and women got equal parental leave; it was one where reproductive labour and childcare were shared by all. Without significant social change, artificial wombs would simply be co-opted into the existing, limited and prejudiced world.

Approaching artificial wombs as a means of relieving inequitable distribution of reproductive labour between men and women also erases the fact that women have never been a monolith. Inequity between different communities of women is as egregious as gender inequity. Social investment in ectogenesis cannot be considered a moral imperative when, as we explored in Chapter 4, global and racialised disparities in care for pregnant people and infants remains so stark. It is certainly true that many women experience pressure to become mothers. The chorus of family members wondering when you might have a baby, the assumption that you must be anxious to find a partner in order to have a child before it is 'too late', and the invasive questions about your plans for children when you do have a partner are familiar tropes.

But Kendal, Smajdor and Firestone alike imply that women are collectively and constantly subjected to pronatalism and are shamed if they dare to be childless. Even as Firestone railed against the social pressure placed on all women to become mothers, however, Black and Indigenous women, women with disabilities and low-income women were fighting against a system that treated them as unworthy of motherhood. As the reproductive justice leader Loretta Ross has emphasised, for these groups of women, forced sterilisation, pressure to use birth control, and the threat of having children removed from their care were equally, if not more, pressing. The constraint that white women like Firestone felt on their reproductive autonomy primarily took the form of being made to feel that motherhood was their duty. But as we considered in Chapter 3, many others were struggling to access the right to have their pregnancies respected and to mother their own children in safety. The idea that ectogenesis could 'free' women from the pressure to reproduce still derives from a narrative that puts white women at the centre, by presuming that all women are oppressed primarily through these means.

Firestone's lack of nuance when it came to how the reproductive experiences of white, middle-class American women differed from those of Black and Indigenous women, disabled women and low-income women within the US is the origin story of this particular feminist approach to ectogenesis. The idea that artificial wombs should be pursued on the basis that they might 'free' women from the experience of pregnancy derives from a privileged position, in which basic needs can be presupposed. It starts from a place where the next step for reproductive freedom is creating an alternative to gestation altogether, rather than securing access to the forms of reproductive care that many relatively privileged women can, for the most part, assume they will receive.

In the 1990s the medical sociologist Dorothy Roberts wrote about the ways that *in vitro* fertilisation had been framed as a new reproductive 'choice'. She commented that this way of understanding reproductive technology as a tool for liberation 'operates like blinders that obscure issues of social power that determine the significance of reproductive freedom and control. It obscures them, not by ignoring them altogether, but by claiming to achieve individual freedom without the need to rectify social inequalities.'[7] The same pattern reverberates in arguments that artificial wombs should be pursued as a means of 'freeing' women from the inequitable burden of pregnancy. If we want to redress the dangers that pregnancy can pose and the uneven weight of childcare responsibilities placed on women, then we can't focus on the pursuit of ectogenesis as a new alternative to pregnancy. Instead we need to focus on redressing the inequities that we looked at in the previous few chapters. We need first to get to a place where racism does not place Black women at risk of premature birth. We need to get to a place where no pregnant person is treated as unworthy of motherhood due to their sexual identity, abilities, ethnicity or class.

The idea that artificial wombs will end inequity between men and women also starts from the assumption that gender is binary. According to this logic, women are defined by a biological capacity to become pregnant, and men are distinct from women because their biology allows them to reproduce without carrying their children. On that basis, the great inequality between genders is understood to be that men cannot gestate and women can. And if these are the agreed-upon facts of life, it makes sense that artificial wombs would offer a solution. By allowing gestation to occur outside the body, the technology would create a new possibility whereby men could take responsibility for gestating a child. But this is an essentialising approach to sex and gender. There are more than two sexes. There are men who can become pregnant and have carried their children. There are women who cannot become pregnant, who are still women. The sweeping proclamation that ectogenesis could present a solution by undoing the association of pregnancy with women's bodies alone erases trans men, non-binary people and gender-queer people who are already gestational parents. It speaks to the tenacity of the narrow, binary understanding of gender that has prevailed in Western culture.

The belief that there are only two sexes is so entrenched that it is imagined that to escape a status quo where women alone are responsible for pregnancy we must create a new technology that would remove gestation from the body completely. This simply is not true. Reproductive biology is not really the 'tyranny' that Firestone believed it to be. The real tyranny is our inability to relinquish archaic ideas of sex and gender. Outside cisgender, heterosexual relationships, people have always created families that eschew binary ideas of 'male' and 'female' parenting roles. The 'tyranny' is social, legal and political: it is the institutions that continue to reinforce reductive, exclusionary ideas of sex, and that enforce narrow definitions of who can be a mother, father or parent. Feminist thinking that reduces 'women' to those who were assigned female sex at birth, and assumes that women are fundamentally 'oppressed' by virtue of their reproductive biology, is part of the problem. It compounds the limiting gender roles that it supposedly critiques. The vitriol of transphobia from self-proclaimed 'gender-critical' commentators that has gripped feminism in Britain is a notable example of this effect in action.

In 2018 the UK government finally followed through on a long-awaited consultation to revise the requirements of the 2004 Gender Recognition Act (GRA). Under the terms of the Act, people seeking to have their gender identity affirmed must undergo the outmoded and unjust process of securing a psychiatric assessment diagnosing them with gender dysphoria, a report detailing their medical treatment, evidence of having lived in their 'acquired gender' for two years and, if they are married, consent from their spouse. These requirements medicalise and stigmatise the experiences of trans people. The Act also fails to recognise non-binary identities and does not extend to people under eighteen. In the lead-up to the consultation there was hope that these restrictive requirements would be removed and the government would shift to a more progressive model, in keeping with countries like Ireland, Norway and Argentina, which allow for self-declaration, a step towards granting people determination over their own gender. This hope was justified, given that many of those who responded to the consultation supported removing the restrictions of the Act.

In the end, the consultation was marred in large part by supposedly feminist 'gender-critical' organisations, such as Fair Play for Women and Women's Place, groups that operate under the impression that allowing people to self-identify will undermine 'sex-based' rights. Liz Truss, then minister for women and equalities, who met each of these groups, was later found to have made misleading statements about having also met trans-rights advocates. She would announce that the Act would stay as it was, aside from a slight reduction in administrative costs for those seeking to legally affirm their gender. This takedown of the consultation by extremely organised associations claiming to be acting in the interests of women speaks to a much broader problem. Their arguments boil down to an investment in the idea that reproductive biology determines one's place in the world. Anyone with organs that led them to be designated male at birth is a man, and men are fundamentally oppressors there-fore, trans women pose a risk of violence to cisgender women in single-sex spaces and, even if they do not, men will claim to be trans just to enter these spaces and abuse women. And those with reproductive organs designated female at birth are women, are fundamentally oppressed, and are always at risk.

The logic here should sound familiar. In Chapter 3 we looked at the way scientific racism was used to justify eugenics, to create hierarchies of human worth. The logic of white supremacy dic-tated that race and ability determined a person's intelligence, capabilities and place in the world. Transphobia follows the same pattern of violent categorisation. It reduces humanity to two categories, male and female, and assumes that who you are and how you move through the world is entirely produced by which of those categories you can be placed into. A world where having a penis determines that you can only ever be a violent predator, and where having a vagina means that you are inevi-tably a victim.

Transphobic feminists insist that not being able to associ-ate womanhood with biological sex would undermine women's rights. What is especially cruel about this idea is that trans women – and Black trans women in particular – are at extremely high risk of experiencing many of the forms of violence that these transphobic groups fear they would exact upon women, such as domestic violence and sexual abuse. This is not about protecting women, but about exerting control over who can be understood as a woman. After the GRA consultation opened, hate crimes against trans women quadrupled over the five years that followed, even as TERFS (trans-exclusionary radical feminists) proclaimed that protecting the rights of trans people would expose cisgender women to harm. The reductive idea that reproductive biology determines someone's sex and subsequently, whether they are an oppressor or the oppressed, is the real tyranny that limits what is possible in redistributing the work of carrying and raising a child. We do not need ectogenesis to disentangle the association of gestation with women, and women alone. Alongside protecting and valuing care work, we need to address social and legal unwillingness to protect fathers, non-binary people and genderqueer people.

In 2020 a British journalist appealed to the Supreme Court in a case that might have been groundbreaking for trans parents. Freddie McConnell is his son's gestational parent and sought legal recognition as a father. While the court recognised and acknowledged that Freddie was a man, it ruled that he could not have legal status as the baby's dad. Instead, for legal purposes, he was a 'male mother'. The court's decision speaks to the rigidity of English law in regarding gestating a baby as something that only a mother can do. This in turn limits the possibilities for removing the association of pregnancy with women's bodies. In effect, the court was saying: we know that you are a man, and we therefore know that men can be pregnant; but only mothers can be responsible for gestating, and therefore you cannot be a father. Acknowledging Freddie as a father who had gestated his child would have meant a precedent for pregnancy as something that a mother, father or any parent could do. That is precisely the kind of vision that Firestone, Smajdor and Kendal put forward: a future where any person of any gender could be responsible for gestation.

Are boundaries around gendered parenthood necessary in

law? It is possible to protect the rights of both parents and children without enforcing binary roles on families. A piece of legislation that passed in 2017 in the Canadian province of Ontario, the All Families Are Equal Act, which was created in consultation with impacted parents, goes much further towards allowing pregnancy to be treated as something that someone of any gender could do. Under the Act, the parent who gives birth has no more legal rights to the child than the parent who does not. Up to four parents can be recognised from birth, and they can choose to be identified as mothers, fathers or simply parents. It is possible to acknowledge and protect the existence of families that do not conform to nuclear models.

We do not need a means of automating gestation to 'degender' pregnancy. We need a substantive undoing of the medical, legal and social practices that undermine the pregnancies and parenthood of people of all genders. The barrier isn't the limit of our bodies. It is institutions that police gender and send the message that a family requires a mother, who carries a pregnancy, and a father, who does not. We have had the tools to escape 'the tyranny of reproductive biology' for a long time: allowing people to self-determine their own genders and protecting access to pregnancy care and parental rights for gestational parents who are not women.

We should reject the false idea that only women gestate. And even as we do so, it is also worth asking whether there is something about *human* gestation that cannot be automated. That doesn't mean that the person who gestates a baby has to be a mother, or that families built through surrogacy or adoption, where a gestational parent is not necessarily involved in a child's life, are missing something.

In the 1860s Tarnier imagined that he was on the cusp of replicating the second half of gestation outside the body and that full ectogenesis was merely a matter of time. In 1923 when J. B. S. Haldane pondered the future of science, he was certain that ectogenesis would replace pregnancy within forty years, but despite constant predictions of its inevitable arrival, we haven't yet achieved this feat. And while one explanation is that there is a lack of appetite to do so, it may also be the case that in the end gestation cannot be replicated. It is possible, of course. And it is possible that we will find that artificial gestation is in every way equivalent to human gestation. That does not mean that we would all choose it. And the first children born from artificial wombs – whose gestational parent, for all intents and purposes, would be a machine – might become adults who decide that they would not choose this, either. If ectogenesis is ever to be empowering to people, to provide them with an additional tool for control over their reproductive lives, then it must start from a place where pregnancy isn't seen as a problem.

Epilogue

To Carry

I did not expect to write this book while pregnant. In fact, but for Covid-19, the whole project might have been over and done with by the time I muttered, 'Wait – what?' at a pregnancy test and ran downstairs to ask my partner if he also saw the little line that I was surely hallucinating. Over the past nine months I have existed in a peculiar space between expert and novice, between intellectualising and experiencing. For years I have researched pregnancy, abortion and reproductive-health law. In the midst of writing about how states regulate our bodies, and how paternalism in medical institutions impacts birth experiences, gestating was suddenly happening to me.

Transphobic 'mommy bloggers' rally around the idea that only 'real women' can be pregnant, that the capacity to gestate a baby, birth it and feed it with your body is inherently and naturally associated with the power of womanhood. This protectionism around pregnancy and motherhood as the defining features of 'femaleness' belies the fact that no matter your gender, pregnancy is profoundly strange. It is the transformation of everything that is familiar about your life, body and mind into something that is unfamiliar. Just because some of our bodies can 'do' pregnancy does not mean that we experience it as a 'natural' state. Nothing about this season of my life has felt like my body doing what it is 'innately' meant to do. On the contrary, it has been a series of dissonant sensations, landing like waves and taking me along for the ride.

What could be more peculiar, and alienating – even if beautiful – than every aspect of your body changing by the day? After years of being utterly unable to eat dairy, here I was, gobbling entire half blocks of Cheddar. That was my first clue: not only did I want cheese, but suddenly my digestive system knew what to do with it. Food that I have always loved became disgusting to me, and simultaneously I longed for meals I'd never eaten. Each morning as I climbed out of bed, the increasing strain on my ankles conveyed that I'd once again grown overnight. Having never had freckles, I've been covered in them since partway through my second trimester, alongside the alteration of the shape of my face itself. I have marvelled at the way I could cut my nails at the beginning of the week, only for them to be back to their previous length by week's end. I have been so delighted by my increased production of the hormone relaxin allowing me to do stretches that had always been out of reach for me, quite literally, that I ignored advice about not overdoing it. This creature that is me, pregnant, is wildly weird.

In the early days when I continued to look exactly like myself, if a little more peaky, I merely had to share that I was pregnant to lose contact with acquaintances who did not want to have children, or to transform from being perceived as a capable, direct, funny adult into a tall baby inundated with insightful questions like 'Are you waddling yet?', 'Is he feeding you?' and 'Are you getting so big?' Who was this person that healthcare providers called 'mama'? Not me. While friends squealed in excitement and proclaimed they could not wait to hold my baby, my head spun at the very thought that there *was* a baby. In my body, so far, there was barely the idea of a baby. Before they would be real and present in the world, I would have to grow them and give birth to them, and change irrevocably who I was in the process. Whether I lost them, whether we made it to term, this would always be something that had happened to me. To become pregnant – to gestate – is to set off on the most profoundly unusual relationship.

In high school, when I was all elbows, the only class I truly loved was English Literature. One afternoon our teacher put Sylvia Plath's 'Metaphors' on the projector. 'What is she describing?' There was a pause, before it clicked for me. Pregnancy. As a roomful of sixteen- and seventeen-year-olds, we agreed that she must have hated being pregnant. The poem was suffused with dread. An elephant? A cow in calf? A train that you've boarded and can't get off? It is only now that I find humour and beauty in that poem: the odd sensation of feeling my baby's head under my belly like a firm apple, my legs retrained into improbable tendrils that still somehow permit me to balance. You can be excited to meet this person that you are growing and simultaneously terrified of all that can go wrong, and of all the ways that you are opening a chapter without knowing how it even starts.

There are, I think, too many cute euphemisms for pregnancy: 'bun in the oven', 'expecting', 'knocked up'. But to gestate means to carry, and for me, this is the one word that does capture how it feels. To 'carry' is a verb. You are continuously engaged in this action throughout your pregnancy. You are carrying the physical weight of a baby – a presence that is more and more pronounced with each passing day. You are carrying the weight of your own emotions, your expectations, hopes and fears, formed by your identity, your past experiences, your family history and your own particular circumstances.

You are also carrying so much more. Social ideas about pregnancy and birth are writ large on your body. You merely need to have the audacity to leave your house with a visible baby bump to become a talking point: the social rules that apply to non-pregnant people do not apply when you are gestating. When I ran down the street on the first sunny day, already feeling wobbly on my swollen feet, a man pointed directly at my stomach and shouted, 'Wow!', only to be followed a few strides later by a woman turning to her friend and saying loudly, for my benefit, 'Here comes mama!' Strangers will see fit to tell you that having a baby is going to be 'all worth it', while simultaneously informing you, in case you did not know, that birth is painful and you should get ready, because parenting is no walk in the park, either.

We live in a world where people continue to barely speak about the realities of what pregnancy physically entails and how long recovery after birth can take until someone is already pregnant. It is when you are carrying a baby that everyone comes out of the woodwork to comment on whether you are doing this whole pregnancy thing right. Choices in pregnancy and post-partum are casually discussed as though they are straightforward: are you going to breastfeed? What we don't talk about enough is that breastfeeding is hard. What we don't talk about enough is the immense pressure many people will face, immediately after birth, to 'choose' breastfeeding; and the immense pressure others will face to 'choose' formula when they are struggling. Pregnant people are 'so big' or 'too small', they eat too much, they don't eat enough, they exercise too much and not enough, they are too old and too young, and everyone from their actual doctors to colleagues and passers-by has something to say about it.

When you carry a pregnancy, you also carry the weight of legal and medical regulations and recommendations, encircling your movements through the world. And you carry the burden of blame when something goes wrong. When the Covid-19 vaccine was first introduced, several of the countries where it was available (including the UK) exercised the precautionary principle when it came to pregnant people. Guidance was released stipulating that the effects of the vaccine in pregnancy and breastfeeding were unknown. Both the Joint Committee on Vaccination and Immunisation and Public Health England advised that those who were pregnant or planning to become pregnant should wait to get vaccinated. Later the advice shifted to permit high-risk pregnant people to get the vaccine. Later still, pregnant people were offered a dose with their age cohorts. It took months before vaccination was actively recommended in pregnancy. And it would take almost a year – during which it became clear that unvaccinated pregnant people were at high risk of severe Covid-19 causing complications for both them and their babies – before pregnant people were prioritised.

When this policy change finally happened, it was alongside a slew of newspaper articles effectively decrying those who were pregnant and breastfeeding for not coming forward. Didn't they know how terribly dangerous it was for them to remain unprotected? Did they not care about their health or their babies? There was little accompanying discussion of the way the messaging had shifted over time, such that people had been told both that it remained unknown whether the vaccine was safe for them and that having Covid-19 while unvaccinated and pregnant could kill them. There was also little accompanying discussion of the numerous anecdotal reports shared among pregnant people themselves, about healthcare providers at clinics advising against vaccination, giving reference to the old recommendations. This example is part of a consistent pattern, wherein pregnant people are expected to sift through often conflicting and changing information about what constitutes a risk and are left to grapple with the consequences alone if something goes wrong either way.

At best, the scrutiny of pregnant people's bodies and the social norms that occur around their activities are felt as mere frustrations and inconveniences. Perhaps you must decide whether to take paracetamol while ill and find yourself wading through contrary recommendations. Or perhaps multiple different reliable sources give you advice about how best to manage exercise, diet and sleep, increasing your anxiety and leading to some wakeful nights. All in all, these experiences may be irritating, but are mitigated by the pleasures of feeling your baby kick, or by the comfort in knowing that you have a solid support system and a healthy pregnancy.

At worst, though, the hyper-regulation of pregnant people's bodies, and the policing of who is considered worthy of carrying a pregnancy, can lead to criminalisation, harm and death. In many parts of the world, laws demand that when an embryo implants and the process of gestation begins, pregnancy must continue. People have carried unwanted pregnancies under these laws and they have also refused them, seeking out the means of termination on their own and confronting stigma and criminal charges as a consequence. In Missouri, a bill proposed in spring 2022 would have all abortion banned from six weeks, including ectopic pregnancies, which begin outside the uterus and are never viable. No living baby can develop from ectopic pregnancies, but they can kill the pregnant person. The statute, like so many others that have preceded it and will follow it with the fall of Roe, sends a clear message: if you are carrying a pregnancy – even if you don't want it, even if it is a death sentence – you must stay that way. Simultaneously, in some of these very same states where laws are passed in an attempt to enforce pregnancy, people who dare to seek help for drug or alcohol dependency while carrying a wanted child face the threat of being detained and criminalised.

The expectations, fears and judgements that pregnant people carry are never distributed evenly. They are racialised, classed and gendered judgements. Black and Indigenous women in North America continue to be subjected to increased surveillance of their choices during pregnancy and to life-threatening racism before, during and after birth. In much of the world, trans masculine, non-binary and genderqueer parents who carry their children face stigma, social scrutiny and barriers to care. Many pregnant people in wealthy nations are monitored to the point of overwhelm, weighed at each medical visit, sent for regular tests to rule out every possible complication (no matter how improbable) and pressured to strictly regulate every aspect of their lives to optimise the health of their babies. Meanwhile in the rest of the world, pregnant people and their babies die in high numbers from easily detectable and preventable causes because of a lack of access to basic resources.

We require gestation for the reproduction of our species. Yet we expect pregnant people to carry so much. It is no wonder that ectogenesis holds such allure. And it will be no wonder if we find that, no matter how far scientific research advances, gestation cannot be replicated. It is a reduction of the physical, emotional and mental work done by pregnant people, of the way they make the best decisions they can for the fetus they carry, even in the absence of any support, to assume that gestating could be automated. It is also a reduction of the immense pressures laid on pregnant people to assume that the weight of these burdens is not so dire as to explain why ectogenesis might be appealing. People get dreamy and fearful about artificial wombs for the same reason. It is compelling to imagine the technology because, taking stock of all the frustrations, expectations and dangers that can come with pregnancy, it allows us to ask: what if none of us had to do this? What if we could simply share this weight, or give it away? And it is frightening, too, because we can look at the harms and injustices that punctuate the experience of gestating in the absence of the technology and envision how it could be used as a new tool for a long-standing practice of co-opting, punishing and controlling the behaviour and bodies of pregnant people.

Could a technology carry all of this? The aspects of gestating that elide the picture so often painted of an autonomous artificial womb are the least tangible ones and are also perhaps the most crucial. The pregnant person mediates the space between the world and the growing baby. Before it is fully formed, they make choices about its care. Before it is born, whether they intend to parent it or not, they navigate the relationships that will lay the foundation for how it will be loved. Pregnancy and birth can be incredible experiences. In the 1920s Vera Brittain observed that just because ectogenesis might come to exist did not mean that anyone would want to use it. It is just as true today that many people don't simply want to have a child, they want to gestate that child. People can go through incredibly challenging emotional and physical experiences to become pregnant. And for many, the process of pregnancy is deeply moving, joyful and affirming.

Perhaps the best possible way to use ectogenesis is as an invitation to confront precisely how much we expect pregnant people to carry, and how little we acknowledge its weight. It might only be through first acknowledging that human gestation is meaningful, such that it might not be possible to automate it, that we will truly initiate the social changes we so desperately need. What kind of world do we need to build before we can really dream about ectogenesis; before it becomes possible to imagine a means of external gestation that might be used in ways that facilitate collectivity and care, rather than harm?

Picture a cluster of five supersized round red balloons, floating upwards with clear, thin cords drifting toward the ground. This is the display you would have seen had you attended the 2018 Reprodutopia exhibit at Dutch Design Week in Amsterdam. Hendrik-Jan Grievink of the Next Nature Network (something of a modern-day Heretics Society) and speculative designer Lisa Mandemaker were invited to create the installation by Dr Guid Oei of Máxima Medical Centre, where contemporary artificial-womb research is ongoing. As visitors would be reminded, the red balloons were not a functional prototype. They were instead created to prompt reflection and debate. At the planning stage, Mandemaker examined graphs provided by scientists, and images of patents from the early research on artificial wombs conducted in the 1950s and 1960s. These metal warmers and machines were disconcerting: cold, distant and inhuman. What form could ectogenesis take that might evoke a different kind of response?

Mandemaker wanted to create a sense of things growing in nature rather than in a lab. She thought of a 'nursery of the future', inspired by groves of new trees, pumpkin patches and botanical gardens, and set about attending to each detail to try to give the finished product a sense of intimacy. What if, instead of being something you would see only in a hospital, the artificial womb was like any other object that you might encounter in your own home? To create unique patterns for each balloon, she generated squares of fabric prints, using sandwich bags filled with liquids sourced from her kitchen, such as jam and sparkling water, which she placed flat under a scanner. The installation was meant to keep open the question of how it might be used, to encourage viewers to express their thoughts. Could it be placed in somebody's living room? The absence of monitoring systems and wires indicated that it could. Could it be operated by a midwife? There was nothing in the design to suggest it could not. The red bulbs perched on slender stems offer another way of conceptualising the artificial womb. Without being too prescriptive, images of the display hint at a different possible path for ectogenesis. One where people might be able to use artificial wombs themselves, in places of their choosing. One where feminist researchers might create a technology to care for pregnant people and neonates that is not sinister or alienating, but familial and empowering.

In a world where reproductive justice was realised, could the artificial womb be used to support pregnant people's self-determination rather than to undermine it? Could people choose to gestate their babies through ectogenesis, to share gestation with a partner, a friend or their community? Could they transfer responsibility for a fetus they did not wish to carry, or use an artificial womb to continue to gestate if pregnancy became dangerous or unsustainable for them?

The precondition for these possibilities is a world where the pregnancies of people of all genders are respected, where reproductive care – from abortion, to contraception, to antenatal care – is available to everyone on their own terms. It is one where eugenic thinking has been abolished and people are granted the means to sustain the families they build, of their own choosing. We cannot yet imagine the possibilities for ectogenesis, for the disobedient future of birth, in our own untenable present. But we might imagine another kind of artificial womb, in another kind of future – a future in which the weight of carrying a pregnancy is shared, not through automating gestation, but through the provision of resources, support and care.

Acknowledgements

Signing a contract to write this book was the bright spot in the bleak and frightening spring of 2020. Writing it was a calm port in a storm that went on longer than I could have imagined. I am wildly grateful to everyone who had a part in making this happen. Thank you to everyone at Profile Books, Wellcome Collection and Aitken Alexander. Without Chris Wellbelove, my agent, this book would not exist. Thank you for being excited about this story before it was on the page. Ellen Johl, you are calm, insightful, rigorous and thoughtful in your feedback. I could not have dreamed up a more supportive editor. Thank you for doing this with me, and for taking all the changes from start to finish in your stride.

Thanks are due also to my colleagues, Julie, Joanna and Angela, whose creativity in dreaming of a better future buoyed me and reminded me of the possibilities to be found in speculative thinking. Thank you to the people who made time in the thick of many Zoom calls to chat with me as I synthesised my ideas, especially Farah Diaz-Tello, Lisa Mandemaker and Clare Murphy. Thank you to Kristin, Elizabeth and Stacey: your early enthusiasm allowed me to imagine that I could write a book before I actually thought I could.

Thank you to my family, near and far, for giving me an unshakable foundation. Finally, thank you to Nathan, Jack and Lydia: I love you so much, and I'm glad every day that you love me.

Notes

1 Of Incubators, Orchids and Artificial Wombs

1. Shulamith Firestone, *The Dialectic of Sex* (New York, William Morrow and Company, 1970), p. 198.

2 An Artificial Foster-Mother

1. Emily A. Partridge, Marcus G. Davey, Matthew A. Hornick, Patrick E. McGovern et al., 'An extra-uterine system to physiologically support the extreme premature lamb', *Nature Communications* 8 (2017), p. 10.
2. Haruo Usuda and Matt Kemp, 'Development of an artificial placenta', *O&G Magazine: Prematurity* 21:1 (2019).
3. Jeffrey P. Baker, *The Machine in the Nursery: Incubator Technology and the Origins of Neonatal Intensive Care* (London, Johns Hopkins Press, 1996).
4. Dawn Raffel, *The Strange Case of Dr Couney* (New York, Blue Rider Press, 2019).
5. Claire Prentice, *Miracle at Coney Island* (Michigan, Brilliance Publishing, 2016).
6. Katie Thornton, 'The Infantorium', *99% Invisible*, episode 381 (3 December 2019).
7. Jeffrey P. Baker, 'Technology in the Nursery: Incubators, Ventilators, and the Rescue of Premature Infants', in *Formative Years: Children's Health in the United States, 1880–2000*, edited by Alexandra Minna

Stern and Howard Markel (Ann Arbor, University of Michigan Press, 2002), p. 81.

8. Unno, Nobuya, 'Development of an Artificial Placenta,' *Next Sex: Ars Electronica*, edited by Gerfried Stocker and Christine Schoepf (New York & Vienna, Springer, 2000), pp 62–68.

9. Katharine Dow, '"The Men who Made the Breakthrough": How the British press represented Patrick Steptoe and Robert Edwards in 1978', *Reproductive Biomedicine and Society Online* 4 (2017), pp. 59–67.

10. The papers of Lesley Brown, which consist of fifteen boxes of archival material, were donated to the Bristol Archives by Louise Brown in 2016.

11. Laurence E. Karp and Roger P. Donahue, 'Preimplantational Ectogenesis: Science and Speculation Concerning In Vitro Fertilization and Related Procedures', *The Western Journal of Medicine* 124 (1976), pp. 282–98.

12. Warnock made this comment at the 2016 Progress Educational Trust Conference, 'Rethinking the Ethics of Embryo Research: Genome Editing, 14 Days and Beyond.' For a more in-depth interview with Mary Warnock on the logic of the 14-day rule and law, see Hurlbut et al (note 13).

13. Benjamin J. Hurlbut, Insoo Hyun, Aaron D. Levine, Robin Lovell-Badge et al., 'Revisiting the Warnock Rule', *Nature Biotechnology* 35 (2017), pp. 1029–42.

14. Carlo Bulletti, Valero Maria Jassoni, Stefania Tabanello, Luca Gianaroli et al., 'Early human pregnancy in vitro utilizing an artificially perfused uterus', *Fertility and Sterility* 49:6 (1988), pp. 997–1001.

15. Christine Rosen, 'Why Not Artificial Wombs?: On the meaning of being born, not incubated', *The New Atlantis*, Fall 2003, www.thenewatlantis.com/publications/why-not-artificial-wombs

3 Ectogenesis for a Brave New World

1. J. B. S. Haldane, 'Daedalus; Or, Science and the Future', in *Haldane's Daedalus Revisited*, edited by Krishna Dronamraju (Oxford, Oxford University Press, 1923).

2. Angela Saini, *Superior: The Return of Race Science* (Boston, Beacon Press, 2019), p. 124.

3. Victoria Brignell, 'The eugenics movement Britain wants to forget', *New Statesman* (9 December 2010).

4. Anna Diamond, 'The 1924 Law That Slammed the Door on Immigrants and the Politicians Who Pushed it Back Open', *Smithsonian* (19 May 2020).

5. *Buck v. Bell*, 274 US 200 (1927), www.supreme.justia.com/cases/federal/us/274/200/#tab-opinion-1931809

6. Harriet A. Washington, *Medical Apartheid: The Dark History of Medical Experimentation on Black Americans from Colonial Times to the Present* (New York, Doubleday, 2007).

7. J. B. S. Haldane, 'Daedalus; Or, Science and the Future', in *Haldane's Daedalus Revisited*, edited by Krishna Dronamraju (Oxford, Oxford University Press, 1923).

8. Anthony M. Ludovici, *Lysistrata; Or, Woman's Future and the Future Woman* (New York, E. P. Dutton & Co., 1924).

9. Norman Haire, *Hymen, Or the Future of Marriage* (London, Kegan Paul, Trench & Trubner, 1927).

10. Joanne Woiak, 'Designing a Brave New World: Eugenics, Politics, and Fiction', *The Public Historian* 29:3 (2007), p. 106.

11. Gregory Pence, 'What's So Good About Natural Motherhood? (In Praise of Unnatural Gestation)', in *Ectogenesis: Artificial Womb Technology and the Future of Human Reproduction*, edited by Scott Gelfand and John R. Shook (New York, Rodopi, 2006), p. 82.

12. Scott Gelfand, 'Ectogenesis and the Ethics of Care', ibid.

13. Christopher Kaczor, 'Could Artificial Wombs End the Abortion Debate?', *National Catholic Bioethics Quarterly* 5:2 (2005), p. 298.

14. It is worth noting that in 2021 BPAS released a strategy document in which the organisation wrote that while its services were inclusive and it would build pathways 'to meet individual needs', it would 'continue to use the word "women" over "people" [to] continue to campaign effectively for reproductive rights'. This occurred in the context of ongoing transphobia within the UK. Using gender-neutral terms like 'pregnant people', and recognising that trans and non-binary people also need reproductive health services, does not undermine fighting for the reproductive rights of cis women. Gender-inclusive language rejects the idea that one's reproductive organs define their identity and place in the world – a concept that is a product of biological determinism.

15. Lynn Paltrow and Jeanne Flavin, 'Arrests of and Forced Interventions on Pregnant Women in the United States, 1973–2005: Implications for Women's Legal Status and Public Health', *Journal of Health Politics, Policy and Law* 38:2 (2013), pp. 300–43.
16. Marie Carmichael Stopes, *Radiant Motherhood* (New York, G. P. Putnam's Sons, 1920).
17. Dorothy Roberts, 'Reproductive Justice, Not Just Rights', *Dissent Magazine* (Fall 2015).
18. Loretta Ross, 'What is Reproductive Justice?', in *Reproductive Justice Briefing Book: A Primer on Reproductive Justice and Social Change* (Pro-choice Public Education Project, 2007), p. 4.
19. J. D. Bernal, *The World, The Flesh and the Devil* (London, Verso, 2017).

4 Mother Machine

1. Sarah Digregorio, 'Artificial Wombs Aren't a Sci-Fi Horror Story', Future Tense: *Slate* (30 January 2020).
2. The CDC uses 'American Indian' and 'Alaska Native' as identity categories in tracking health outcomes for neonates and pregnant people, but there are more than 500 federally recognised tribes in the US. One failing of tracked health data is this lack of specificity, both in terms of identifying which specific groups are most impacted and of using precise and self-determined language.
3. The World Health Organization, Fact Sheet: 'Preterm Birth', 19 February 2018.
4. Jeffrey D. Horbar, Erika M. Edwards, Lucy T. Greenberg et al., 'Racial Segregation and Inequality in the Neonatal Intensive Care Unit for Very Low-Birth-Weight and Very Preterm Infants', *JAMA Pediatrics* 173:5 (2019), pp. 455–61.
5. Harriet Washington, *Medical Apartheid* (2007).
6. Dána-Ain Davis, *Reproductive Injustice: Racism, Pregnancy, and Premature Birth* (New York, NYU Press, 2019), p. 102 (both quotes).
7. Usuda and Kemp, 'Development of an artificial placenta' (2019).
8. Dr Joia Crear-Perry, 'The Black Maternal Mortality Rate in the US Is an International Crisis', *The Root* (30 September 2016).
9. Partridge, Davey, Hornick, McGovern et al., 'An extra-uterine system to physiologically support the extreme premature lamb' (2017).
10. Dorothy Roberts, 'Reproductive Justice, Not Just Rights' (2015).

11. Loretta Ross, Lynn Roberts, Erika Derkas, Whitney Peoples and Pamela Bridgewater Toure, *Radical Reproductive Justice* (New York, The Feminist Press, 2017).

12. Dr Joia Crear-Perry, 'The Birth Equity Agenda: A Blueprint for Reproductive Health and Wellbeing', National Birth Equity Collaborative (16 June 2020).

13. Andrea Nove, Ingrid K. Friberg, Luc de Bernis, Fran McConville et al., 'Potential impact of midwives in preventing and reducing maternal and neonatal mortality and stillbirths: a Lives Saved Tool modelling study', *The Lancet Global Health* 9:1 (2020).

5 The Abortion Solution

1. Mark A. Goldstein, 'Choice Rights and Abortion: The Begetting Choice Right and State Obstacles to Choice in Light of Artificial Womb Technology', *Southern California Law Review* 51:5 (1978), p. 880.

2. Kimala Price, 'What is Reproductive Justice?: How Women of Color Activists Are Redefining the Pro-Choice Paradigm', *Meridians: Feminism, race, transnationalism* 10:2 (2010), p. 46.

3. Emily Jackson, 'Degendering Reproduction?', *Medical Law Review* 16:3 (Autumn 2008), pp. 346–68.

4. Sarah Langford, 'An End to Abortion? A Feminist Critique of the "Ectogenetic Solution" to Abortion', *Women's Studies International Forum* 31 (2008), p. 267.

5. Emily Jackson, 'Abortion, Autonomy, and Prenatal Diagnosis', *Social and Legal Studies* 9:4 (2000), p. 469.

6. Peter Singer and Deane Wells, 'Ectogenesis', *Journal of Medical Ethics* 9:192 (1983), p. 12.

7. It bears noting that much of the language of abortion law constructs disabilities as an exception – in other words, it implies that to have a child with a disability would be so significantly damaging to a person as to justify them having an abortion where the law would otherwise not permit it. This also has the effect of further stigmatising people with disabilities and people who choose to carry on with their pregnancies when screening suggests difference. Part of ending the criminalisation of abortion is ending such language of exception. Just as no one should have to provide a justification for seeking abortion, no one should be made to feel that they must justify their choice to

continue their pregnancy or that a life with disabilities is not worth living.

8. Chris Kaposy and Jocelyn Downie, 'Judicial Reasoning about Pregnancy and Choice', *Health Law Journal* 16 (2008), p. 290.

6 The Tyranny of Biology

1. Shulamith Firestone, *The Dialectic of Sex* (1970), p. 198.
2. Shulamith Firestone, *The Dialectic of Sex: The Case for Feminist Revolution* (London, The Women's Press, 1979).
3. Anna Smajdor, 'The Moral Imperative for Ectogenesis,' *The Cambridge Quarterly of Healthcare Ethics* 16 (2007) pp. 336–345.
4. Evie Kendal, *Equal Opportunity and the Case for State-Sponsored Ectogenesis* (Basingstoke, Palgrave Macmillan, 2015).
5. Gregory Pence, 'What's So Good About Natural Motherhood?', in Gelfand and Shook (eds), *Ectogenesis: Artificial Womb Technology and the Future of Human Reproduction* (2006), pp. 77–88.
6. Evie Kendal, *Equal Opportunity and the Case for State-Sponsored Ectogenesis* (2015), p. 12.
7. Dorothy E. Roberts, *Killing the Black Body: Race, Reproduction, and the Meaning of Liberty* (New York, Vintage Books, 1999), p. 298.

Index